Water Meadows

History, Ecology and Conservation

edited by
Hadrian Cook and Tom Williamson

WIND*gather*
PRESS

Frontispiece
View across the Seven
Acres at the Harnham
water meadows,
Salisbury, Wiltshire.

Published by: Windgather Press Ltd, 29 Bishop Road, Bollington,
Macclesfield, Cheshire SK10 5NX

Distributed by: Central Books Ltd, 99 Wallis Road, London E9 5LN

British Library Cataloguing-in-Publication Data
A catalogue record for this book is available from the British Library

ISBN 10 1-905119-12-7
ISBN 13 978-1-905119-12-7

Designed, typeset and originated by Carnegie Publishing Ltd,
Chatsworth Road, Lancaster

Printed and bound by Cambridge Printing, Cambridge

Contents

List of Figures

v

Acknowledgements

The chapters in this book are based on papers delivered at a conference organised by the Friends of the Harnham Water Meadows Trust at the Salisbury and South Wiltshire Museum on 26 and 27 March 2004, and we would like to thank the Friends of the Trust, and in particular Tim Tatton-Brown and Jennifer Bowen, for all their help and encouragement. We would also like to thank all those who have given advice and assistance to the individual authors: in particular, Graham Brown, Bruce Campbell, Michael Cowan, Andrew Lynn, Clare Tawney, Owen Thompson, Eric Viser, Susanna Wade Martins and Robert Wilson North, together with the staff of the Dorset Record Office, the Norfolk Record Office, the Wiltshire and Swindon Record Office, and the archive of the Centre for Ecology and Hydrology. Thanks also to the University of Reading Institute of Agricultural History and Museum of English Rural Life, for Figures 6 and 14; the Centre for Ecology and Hydrology, for Figure 31; Hampshire County Council, for Figure 5; Doncaster Archives and Fiona Cowell, for Figure 23; the Ministry of Defence, for Figure 26; Wiltshire and Swindon Record Office, for Figure 10; Dorset Record Office, for Figure 4; Deborah Overton for Figures 11 and 12; English Heritage, for Figures 3, 8 and 9; Cambridge University Committee for Aerial Photography, for Figures 17 and 35; Bruce Campbell, for Figure 15; Kathy Stearne, for Figures 19 and 20; Dr Eric Visser, for Figure 30; Gary Battell, for Figure 13; the Tale Valley Trust and Exeter Archaeology, for Figure 22; and Phillip Judge, for Figures 1 and 21.

Contributors

Joe Bettey was Reader in Local History at the University of Bristol. He has published numerous articles on the West Country, and is particularly interested in the history of farming and rural society. He has made a special study of the development of water meadows in Wiltshire and Dorset.

Hadrian Cook is Development and Education Officer for the Harnham Water Meadows Trust in Salisbury, UK. Before that he taught and researched in soil science, hydrology and historic landscapes in the University of London for eighteen years. He specialises in catchment management, concentrating on the floodplain.

Ian Cummings currently teaches ecology at Norwich City College (Regional College of UEA) and has worked with English Nature and the voluntary sector. His research interests are primarily in the conservation management of ancient woodlands and water meadows, and field studies; current work focuses on the plant ecology of bedwork water meadows.

Roger L. Cutting has written widely on the subject of bedwork water meadows with a particular emphasis on the science of their operation. He is presently a Senior Lecturer in the Faculty of Education at the University of Plymouth, where he teaches on a range of courses relating to environmental science and environmental education.

Andrew Fielder has worked for Defra's Rural Development Service since 2001, managing The Test and Avon Valley Environmentally Sensitive Area Schemes. Prior to that he was involved in agricultural consultancy work in Wiltshire and Dorset.

Peter Martin learned the art of drowning from his father. He took over the family's farm, including the Britford Water Meadows Site of Special Scientific Interest, in the 1950s and has managed the meadows through good and bad times ever since, making him one of the most experienced drowners in the country.

Kathy Stearne graduated from Reading University with a BSc in Soil Science and has worked in ADAS and Defra since 1980. She was awarded a PhD in 2005 for her work on the management of water meadows, and is also a partner in Green Mark International, an independent consultancy offering consultancy for environmental projects.

Christopher Taylor was formerly head of Archaeological Survey for the Royal Commission on the Historical Monuments of England and is a Fellow of the British Academy. He is author of numerous books on settlement, landscape and garden history, including *Village and Farmstead: A History of Rural Settlement in England* and *Parks and Gardens in Britain: A Landscape History from the Air*.

Tom Williamson is Reader in Landscape History at the University of East Anglia. He has written widely on landscape archaeology, agricultural history and the history of landscape design.

CHAPTER ONE

Introducing Water Meadows

Tom Williamson and Hadrian Cook

The purpose of water meadows

The term 'water meadow' is now rather loosely applied to any area of low-lying land used for producing hay, the dried grass used to feed livestock through the winter. But until relatively recently it had a more restricted meaning: a water meadow was an area of grassland in which the quantity of grass was increased, and its quality improved, through artificial irrigation. At certain times of the year, water was moved on to the meadow via a system of channels, passed in a continuous moving film through the sward, and then removed via a network of drains. Constant movement was imperative, for stagnant (and hence oxygen-deficient) water would damage the grass. Irrigation of this kind – perhaps more suitably described, using the traditional terms, as 'floating' or 'drowning' – was carried out for a number of reasons.* Probably the most important was to secure an 'early bite' for sheep or cattle in the spring. British livestock farmers have always had to contend with the period between October and April when grass grows little, if at all; and especially with the 'hungry gap', the months of March and April, when supplies of hay and other fodder run low. Winter irrigation was intended to shorten the period of time during which sheep or cattle were fed on fodder by bringing forward the growth of the spring grass by several weeks. Irrigation usually began before Christmas and continued until early March. The grass was typically irrigated for between four and six days at a time, the moving water serving to maintain the ground temperature above the critical 5 °C necessary to stimulate growth. The stock were put on to the meadows (which really, at this stage of the year, functioned as pastures) after a short period of drying, and were kept there until late April or May. They were then moved on to other grazing and a second period of irrigation commenced, this time usually by night and of shorter duration. The purpose was now to enhance the summer hay crop, by maintaining, through the late spring and early summer, the moist conditions ideal for grass growth.

These were the two main reasons for irrigation, but a number of others

---

* The two terms are almost, but not quite, interchangeable: 'floating' was often used for this form of management as a whole; 'drowning' seems often to have been restricted to the actual act of applying the water.

are mentioned by seventeenth-, eighteenth- and nineteenth-century agricultural writers, or have been identified by modern scientists. The moving water dressed the sward with lime, usually in the form of dissolved calcium carbonate (in solution as calcium hydrogen carbonate), and with suspended sediment which fertilised the soil (some early writers advocated the use of dirty water, from urban streets or farmyards). It also had complex chemical effects, improving the sward by, in particular, favouring the growth of more palatable grasses over other species, in ways which Ian Cummings and Roger Cutting explain in Chapter Seven.

Types of water meadow

There were two basic types of water meadow although, as we shall see, there were also a number of related, and hybrid, methods of water management. The most important of these basic types, and today the most archaeologically visible, were the so-called *bedworks*, which are still, although now largely in redundant form, a familiar feature of the river valleys of southern England. These were complex and sophisticated systems which were used where wide, level floodplains were to be irrigated (Figure 1). A major watercourse was dammed by a weir, immediately above which water was fed, via a 'hatch' or sluice, into an artificial channel (a 'carriage' or 'carrier'). This ran roughly parallel with the river, but with a lesser gradient, so that when the water reached the area to be irrigated it was flowing at around a metre or more above its natural level. It was then led through smaller carriers and hatches into spade-dug channels which ran along the spines of low ridges, called 'beds' defining

FIGURE I.
The layout of a typical bedwork water meadow.

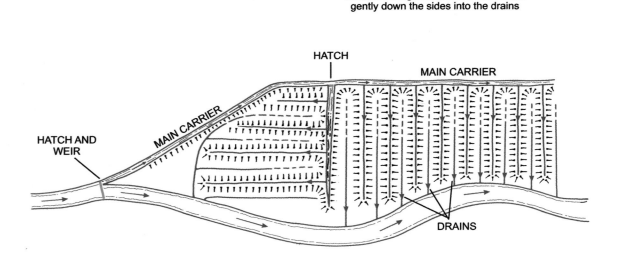

FIGURE 2.
A bedwork being floated at Britford in Wiltshire. The water running along the channel cut into the top of the ridge flows down the side and into the drain in the foreground. In the distance is the main carrier, which brings water to the entire system.

'panes', such that they superficially resemble the ridge and furrow earthworks of former open fields (Figure 2). The water flowed gently down the side of the ridges into drains (sometimes called 'drawns') in the adjacent furrows, and was then – either directly, or via a 'tail drain' – returned to the river. The 'drowner', 'meadman' or 'waterman', the skilled individual responsible for the management of the meadows, controlled the water carefully, irrigating blocks of panes in turn and ensuring an even and steady flow of water, not only by adjusting the hatches or sluices but also by placing 'stops' or turfs in the smaller channels.

Bedworks were, as we shall see, relatively expensive to create, not only because of the cost of raising the beds and constructing the hatches and sluices, but also because culverts and bridges were required to provide access to the meadows for carts when the hay was harvested. Some meadows were also provided with aqueducts to take the carriers across the river in order to reach awkward angles and corners of ground. Bedworks also required the attention of a trained drowner, able to maintain the steady, even flow across the surface of the ground.

FIGURE 3.
Aerial view of a disused
catchwork water
meadow at Lower
House, Cutcombe, on
the Brendon Hill in
Somerset.
© ENGLISH HERITAGE

The second main type of water meadow, the *catchwork* or *catch-meadow*, was rather simpler. Water from a spring or stream was led in a channel or headmain along the side of a hill; it was encouraged to overflow the channel, or passed through openings in its side, so that it spread down the slope before returning to the parent stream or being taken off the land by an artificial tail drain. Sometimes a series of parallel gutters was created, to encourage a more even flow of water down the slope. Catchworks were cheaper to construct than bedworks, although only suitable for hilly terrain. They were also rather simpler to operate, and often did not require the attention of a specialised 'drowner' (Figure 3) (Thomas 1998; Sheail 1971).

4

The geography of floating

Water meadows could, at various times, be found in almost all areas of Britain, but they were most characteristic of southern England – of the valleys of the various chalk rivers in the Hampshire basin and surrounding areas, in the counties of Dorset, Hampshire and Wiltshire. By the eighteenth century almost every significant floodplain in this region was occupied by bedwork systems of varying complexity and, as Joe Bettey explains in Chapter Two, 'floating' formed a central part of the region's farming economy. The reasons for the supremacy of Wessex meadows are complex, as we shall see; they are in part the result of social and economic factors, and in part the consequence of environmental circumstances.

The construction of bedwork water meadows seems, on the present evidence, to have begun in Wessex in the years around 1600. They were principally used to provide an early bite and increase the summer hay crop, although the improvements they brought to the quality of the grass, through oxygenation and the dressing of the sward with suspended nutrients, were also widely recognised. The meadows formed a central feature of the local 'sheep-corn' system of agriculture, in which the cereal yields from the thin, poor soils were enhanced by the manure from sheep flocks, systematically night-folded on the arable land. Not only was irrigation in this region more widely practised than elsewhere; it also continued longer, into the twentieth century, although it was by then being used in rather different ways in the farming economy, as Kathy Stearne explains in Chapter Nine. A few meadows in the region are still regularly floated, as Peter Martin, one of the last irrigating farmers, describes in Chapter Ten.

Outside Wessex the history and use of meadow irrigation was rather different. In other regions water meadows never held the same kind of central role in agriculture, took different forms, and were often valued for rather different reasons. Across much of western England, especially lowland districts in Devon, Herefordshire and Somerset, for example, irrigation was widely practised in the seventeenth and eighteenth centuries but usually in the form of catchworks, rather than bedworks, due to the nature of the topography. Moreover, in most of these areas the increases in livestock numbers made possible by the technique were mainly valued in their own right, rather than as a way of increasing the output from arable land. In the later eighteenth and nineteenth centuries the practice seems to have spread into many upland areas of England, and also into Scotland and Wales. Here, too, irrigation was used to bring on an early growth of grass, and sometimes to improve the hay crop, but the other benefits that water meadows conferred were often of greater importance. In particular, irrigation served to improve the quality of the upland pastures, partly because it mimicked a widespread natural phenomenon. As L. Dudley Stamp noted in 1950, 'It is well known that "flushes" of excellent grass are found round springs on hillsides and are attractive to sheep as green oases in mountainous country covered with heather or other rough grazing' (Stamp 1950, 84). The dissolved oxygen in the flowing water encouraged the growth of white clover and broadleaved grasses such as red fescue at the expense of coarser grasses such as *Nardus* sp; while, in a variety of ways, soil quality was improved. But, in addition, farmers in northern and western regions put particular value on the way that irrigation could be used to dress the sward with lime and other nutrients. Clarke typically advocated the use of dung mixed with water, and the run-off from farmyards, in irrigation schemes in Brecknockshire (Clarke 1795).

Elsewhere in Britain there were other differences of emphasis, reflecting local and regional environmental circumstances and the particular needs of farmers. Some water meadows in the east of England, for example, seem to have been used primarily or exclusively to improve the summer hay crop, rather than to encourage the early growth of grass. Water meadows, in short, were a complex and diverse phenomenon, and while it was unquestionably in Wessex that they achieved their highest form of development and became most important in the farming economy, we need to study their history in other regions in its own terms, and not simply as a pale reflection or imitation of Wessex practice.

The fact that meadow irrigation outside its Wessex heartlands was carried out for such a range of reasons, and took such a variety of forms, means that any search for the 'origins' of floating is fraught with difficulties. The old orthodoxy, that the technique was only invented in the late sixteenth century by a Herefordshire landowner called Rowland Vaughan (whose book *The Most Approved and Long Experienced Water Workes* was published in 1610), is certainly wrong. Forms of meadow irrigation were practised in the Middle Ages and, as Christopher Taylor suggests in Chapter Three, it is not impossible that water meadows existed in Roman Britain.

The importance of water meadows

Water meadows have long been a subject of great interest to agricultural historians and several – most notably, Eric Kerridge and Joe Bettey – have demonstrated their significance in increasing agricultural productivity in the early modern period. Less attention has been paid to the ways in which a lack of suitable opportunities for meadow irrigation (coupled with a relative paucity of meadows of any kind) may have stimulated other kinds of agricultural innovation in the course of the seventeenth and eighteenth centuries, leading to significant regional variations in farming practice – a subject which is examined in more detail in Chapter Four. Water meadows are also a topic of increasing interest to archaeologists, especially those researching the post-medieval rural landscape. Archaeological examination of the remains of bedworks and catchworks can tell us much about their construction, operation and chronology which cannot be learnt from documentary sources, as Christopher Taylor demonstrates in Chapter Three. Moreover, water meadows were, and still are, an important part of the 'traditional' farming landscape.

Although few remain in operation their remains help to convey that 'sense of place', the importance of which is increasingly emphasised in government policy: they are thus a major concern for those working in public archaeology and involved in the curation of the archaeological heritage. Yet at the same time water meadows have a significance in biological conservation, whether they remain operational (as is rarely the case), are now redundant but still managed as grazing land, or are regenerating through natural succession to scrub or wet woodland. At a time when government subsidies to agriculture are shifting away from production and towards conservation, water meadows and their management as parts of the historic landscape are subjects which are rapidly rising up the agenda of public policy.

This short volume is the outcome of a conference held at Salisbury – close to the famous Harnham meadows, and under the aegis of the Harnham Water Meadows Trust and its Friends – in 2004. It examines the subject of water meadows from a wide range of viewpoints, with contributions from archaeologists, historians and environmental scientists, as well as from one of the last active 'drowners' in England. It deals with the history, ecology, management and future of meadows, summarising – in more detail than in any previous volume – what is known about these relics of an almost lost technology. At the same time, however, we hope that this book will point the way for future research, for while we now know much about this fascinating subject, a great deal yet remains to be discovered.

CHAPTER TWO

The Floated Water Meadows of Wessex: A Triumph of English Agriculture

Joe Bettey

The importance of the fold

From the seventeenth century until the beginning of the twentieth century the early grass produced by watered meadows was a vital element in farming throughout the chalklands of Wiltshire, Hampshire and Dorset. Before the influx of cheap grain from across the Atlantic during the last decades of the nineteenth century, the primary aim of chalkland farming was the production of wheat and barley, and these crops could only be grown on thin chalk soils by the use of the sheep fold. The sheep flocks which had fed all day on the downs were close-folded on the cornland at night, fertilising it with their dung. The fold was moved each day to another part of the arable land so that the whole of the fields was eventually covered, and each tenant's portion received the benefit of the dung in turn. The size of the flocks, and thus the extent of the land on which they could be intensively folded, was limited by the number of sheep which could be sustained by hay during the 'hungry gap' of March and April, when the downland grazing was inadequate, and before the new season's grass had begun to grow. The water meadows provided a breakthrough in this age-old barrier to agricultural progress. By carefully controlled watering which covered the grass with a shallow sheet of moving water, protecting it from frost and encouraging growth, water meadows produced a lush supply of grass several weeks before natural grazing was available. Later watering would produce another abundant crop of grass, which would be cut for hay. Thus larger sheep flocks could be maintained, more cornland could be enriched by the dung of the sheep fold, cultivation could be extended and greater crops of wheat and barley could be produced. Eighteenth-century observers described the water meadows as 'the sheetanchor of chalkland husbandry', 'so useful as to be almost indispensable', and their value to farmers as 'almost incalculable' (Claridge 1793, 34; Davis 1794, 34–5; Bettey 1999).

There are numerous documentary references to the importance that farmers attached to the benefits which the sheep flock provided for corn land; these emphasise that sheep were, indeed, primarily valued for the dung they produced. At Wyke Regis, Dorset, in 1642, for example, several tenants were fined for feeding their sheep by day on the manorial downland and then folding them at night on the arable land of a neighbouring manor (DRO P5/MA5 Wyke Regis Manorial Account Book). Likewise, at Milborne St Andrew, Dorset, in 1627, a tenant was fined for feeding his sheep by day 'in the commons of this manor and in the night time doth penn the same … upon other lord's land to the impoverishinge of the Lord's tenants of this manor' (DRO D/RGB 2 Milborne Churchstone in Milborne St Andrew). In a valuation of Sydling St Nicholas, Dorset (part of the estates of Winchester College), made in 1776, the surveyor commented on the sheepwalks on the downs that 'The sheep are kept primarily to produce manure for the Arable Lands, which is the greatest profit gained by them' (Winchester College MSS 21429a). Thomas Davis, who was steward to the Marquess of Bath at Longleat, writing in the *General Survey of the Farming of Wiltshire* for the Board of Agriculture in 1794, described the fold as being composed of hurdles and enclosing up to 1,000 sheep to an acre, and affirmed that 'The first and principal purpose of keeping sheep is undoubtedly the dung of the sheepfold, and the second is the wool' (Davis 1794, 149). The sheep fold, then, was a major reason why the system of common arable fields, with the strips of each tenant scattered throughout, survived for so long in the chalklands, in spite of the 'inconvenience which attended upon it' (Davis 1794, 149). Land on either side of chalk streams, which was occasionally flooded during the winter months, had always been highly prized as a source of hay, as is evident from the values accorded to meadows in the Domesday Survey and in various medieval records and manorial court rolls. Not until the early seventeenth century, however, do elaborate schemes seem to have been introduced for artificially controlled watering, using hatches, weirs, channels and drains (Bettey 1977; Cowan 1982; 2005).

Seventeenth-century floating in Dorset

As is discussed in Chapter Eight, the water from the fast-flowing chalkland streams was ideal for watering meadows, but the construction of a water meadow was nonetheless complex and expensive. The water had to be diverted by means of weirs and hatches constructed in the stream, the meadow had to be levelled and ridges created across its surface with channels, along which the water was made to flow, gently overspilling down the sides of the ridges, or 'panes', in order to cover all the grass. The water was then collected by a network of drains and returned to the river. It was essential that the water was kept moving over the grass, since if it was allowed to stagnate or lie in pools this would produce weeds and rushes, killing the grass instead of encouraging its early growth. Thomas Davis noted that 'The very principle of water-meadows will not allow the water to be stagnant; it must always be kept in action to be of any service' (Davis 1794, 123). The construction and operation of the meadows demanded great expertise, and the 'watermen' or 'drowners' who managed the system and controlled the flow of water became important figures in the farming of the chalkland region.

It is clear that by the early years of the seventeenth century the principles and practical details of controlled watering were already well understood along the Frome and Piddle valleys in Dorset (Figure 4). This was before the publication in 1610 of Rowland Vaughan's book *The Most Approved and Long Experienced Water Workes*, which described his experiments in the Golden Valley in Herefordshire. Definite evidence of a fully developed system comes from Ilsington near Puddletown in 1608, when four of the tenants, Richard Jolife, Walter Jacob, Gregory Hooper and John Herne, agreed to construct a watercourse 900 yards long and 7 feet wide across their tenements, bringing water from the river Frome to their meadows. They also agreed to maintain and scour the channel as necessary and to share the right to use the water as follows:

> Joliffe who undertook the major work to have the water during October, December, February, April, June and August each year.
> Jacob to have the water for the first 15 days of November, January, March, May, July and September each year.
> Hooper to have the water for ten days in November, January, March, May, July and September beginning on the 16th of each month.
> Herne to have water for the last five days of the above months (WSRO 873/85)

The landowner, Henry Arnold, enthusiastically encouraged the scheme and his son later claimed that his father had himself spent more than £200 on the project. Within a few years the Ilsington water meadows were extended 'into and over all or any other partes of or parcells of the meadowes and moores for the bettering thereof' (WSRO 873/85). All the tenants who benefited agreed to contribute to the costs. Further evidence for a fully developed water meadow occurs in the manorial records of Affpuddle, Dorset, in 1605–10. The landlord, Sir Edward Lawrence, was interested in agricultural improvement and from 1606 encouraged his tenants to combine forces to dig the necessary channels and drains and install hatches in the river Piddle to water the area called Eastmeade, in order to produce early grass for the communal sheep flock. In 1610 three men were appointed to supervise the watering of the meadow and the tenants agreed to pay in proportion to the size of their holdings (DRO D/FRA/Ml–2 Court Book of Affpuddle 1589–1612).

The water meadows at Ilsington and Affpuddle were evidently successful, and soon afterwards similar work was carried out on the neighbouring manors of Briantspuddle and Pallington. By 1629 three common meadows at Affpuddle – Northmeade, Westmeade and Eastmeade – were being watered and the manorial court ordered that the schemes should be extended by the appointment of 'a fitt and able man for the worke' and that further channels should be cut 'for the better conveyinge and carriage of the water for the watering

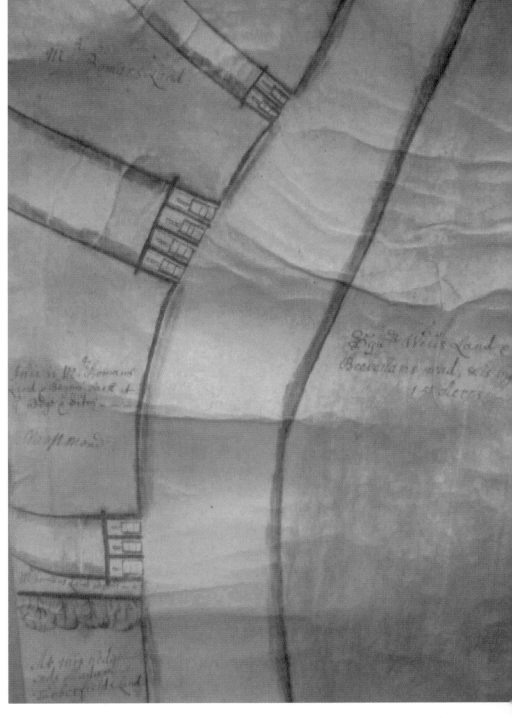

FIGURE 4.
Water meadow hatches
on the river Frome
near Dorchester,
Dorset, shown on a
map of 1675.

of every tenant's meadowe ... and none to interrupt the said workemen in his worke about the carriadge of the said water' (DRO D/FRAJM1–2).

The efficiency with which these water meadows produced early grass and reliable crops of hay soon encouraged similar schemes elsewhere. At Puddletown, 3 miles (5 km) along the valley of the Piddle from Affpuddle, the manorial court book contains references to watering meadows from the 1620s. In 1629 the Puddletown manorial court agreed to turn the whole of the common meadow at Broadmoor into a water meadow. It is significant that the manorial lord, Henry Hastings, was himself present in court when the important decision was taken, although he did not live at Puddletown and the court proceedings were usually conducted by the steward, Richard Russell. Russell was foremost in supporting the scheme, and had evidently already begun work on watering his own meadow land. The agreement arrived at was as follows:

> The honorable Henrie Hastings, esquire, Lord of the same manor being present with the tenants of the same and a greate debate being theare had and questions moved by some of the tenants about wateringe and Improvinge theire groundes and theare heard at large. (DRO D/PUDB1/1/3 Court Book of Puddletown 1624–38)

It was also agreed that Mr Richard Russell and others should be allowed to continue with the work already started 'for wateringe and Improvinge of theire groundes in Broadmoor'. This would involve cutting a new watercourse from the Piddle and constructing weirs, bays, dams and sluices. In this new and untried project success was not assured, and the agreement contained a provision that 'yf yt shall appeare after the maine watercourse shalbe made through the saide grounde, thatt Improvement cannot be made ... then the said Mr Russell shall fill in the said watercourse again at his owne costs' (DRO D/PUDB1/1/3 Court Book of Puddletown 1624–38). The scheme proved to be successful, and soon afterwards agreement was reached in the manorial court to water 255 acres (103 ha), and that the meadow land should be 'parted and divided to them [the tenants] by Lot by proportion according to every man's living' (DRO D/PUDB1/1/3). The ideas soon spread to neighbouring manors. For example, at Winfrith Newburgh a water meadow was created beside a tributary of the Frome, and an agreement is recorded in 1635 between Theophilus, Earl of Suffolk, who was lord of the manor, and twenty-two of his tenants, for the creation of a water meadow at Winfrith Mead. It was agreed that:

> In consideration the said Earle is pleased by way of watering or otherwise at his owne Charge to improve the meadow called Winfrith Mead within the said mannor of Winfrith, wee the freeholders, leaseholders and copyholders of the said mannor whose names are underwritten in consideration of the charge that the said Earle shalbe at about the said Improvement and of the benefitt that shall accrue thereby, doe hereby for us and our successors convenant and promise to and with the said Earle his heirs or assigns the value of one Moiety of the profits that shalbe raised by meanes of the said Improvement, to be valued by two indifferent men. (DRO D/WLC/E130)

Soon afterwards water meadows were created in several other manors belonging to the Earl of Suffolk along the Frome and its tributaries. Other water meadows in the district were constructed on the lands of John Trenchard, who was an active land speculator and money-lender in Dorset. The effect of these developments was to change the appearance of the Frome and Piddle valleys. Writing during the 1630s, a local author, Thomas Gerard, described the Frome passing 'amongst most pleasant Meadows manie of which of late yeares have been by industrie so made of barren Bogges' (Gerard 1980 edn, 64, 75).

Floating in Wiltshire

During the early seventeenth century water meadows were also enthusiastically developed in Wiltshire, especially along the chalk streams that converged on Wilton and Salisbury – the Avon, Wylye, Nadder and Ebble – in a region dominated by the estates of the Earl of Pembroke. Rowland Vaughan, whose experiments with water meadows in Herefordshire have already been mentioned, had family and business connections with the Earl of Pembroke, to whom his book published in 1610 was dedicated. This may explain the Earl's enthusiasm for the development of water meadows on his Wiltshire estates. There are references to the building of hatches and the digging of channels in the manorial court records of several of the Earl's manors during the 1620s, and it is evident that water meadows were already being introduced there. The success of these early projects encouraged the adoption of a major scheme in the manor of Wylye in 1632. At a meeting of the manorial court, presided over by the steward, William Knight, the tenants agreed to share the costs of watering the common meadow. John Knight, of nearby Stockton, was to be employed 'to drawe a sufficient and competent quantitie of water of the River of Wylye out of the same River, sufficiently to water and flott all the said groundes or soe much thereof as by industry and art may be watered or flotted'. Knight was to be paid 14s 0d for every acre of water meadow created, and thereafter 2s 0d per annum for each acre maintained (Kerridge 1953a, 138–40).

The Wiltshire antiquarian John Aubrey, who possessed land in the Ebble valley at Broad Chalke, wrote that the water meadows were created there in 1635, and also described those on the Kennet in north-east Wiltshire in c. 1680: 'The watered meadows all along from Marlborough to Hungerford, Ramsbury and Littlecot, at the latter end of Aprill, are yellow with butter flowers' (Aubrey 1969, 104). Water meadows soon spread into the chalkland valleys of Hampshire (Figure 5), and by the 1650s watering had been introduced along the valleys of the Test, Itchen, Meon and Wey (Bowie 1987). In 1669 John Worlidge of Petersfield could describe the development as 'one of the most convenient and advantageous improvements in England within these few years' (Worlidge 1669), and in a Report to the Georgicall Committee of the Royal Society on the Agriculture of Dorset in 1665, the Dorset landowner Robert Seymer, of Hanford, wrote that 'the greatest improvement they have for their ground is by winter watering it, if it lye convenient for a River or Lesser streame to run over it' (Royal Society, MSS 10.3/10; Kerridge 1954).

FIGURE 5.
Water meadows on
the river Avon in
Hampshire: the
floodwaters pick out
the pattern of beds,
drains and carriers
with particular clarity.
© HAMPSHIRE
COUNTY COUNCIL

The organisation of floating

The rapid expansion of water meadows during the seventeenth century led to the emergence of professional surveyors and watermen, or 'drowners' (Figure 6). When, in 1659, a scheme for making a water meadow at Charlton Marshall, on the river Stour in Dorset, was proposed by the tenants, 'two able and sufficient carpenters' were brought from Tolpuddle to make the hatches, while Henry Phelps of Turners Puddle, 'a known Ancient Able and well Experienced waterman', was sent for to supervise the whole project, 'soe ordering the water whereby that the said groundes might be well watered … as farr as the strength of the River would cover' (TNA:PRO C5/58/15). The Charlton Marshall scheme was encouraged and supported by the lord of the manor, Sir Ralph Bankes of Kingston Lacy, and by the Provost of Eton College which had freehold rights in the meadows (TNA:PRO C5/58/15). In 1630, at the manorial court at Affpuddle, the tenants agreed to employ a full-time 'waterman' to look after the meadows. He was to maintain all the weirs, hatches, channels and drains, while the tenants themselves agreed not to interfere with his work (DRO D/FRA/M4a Affpuddle Court Book 31 March 1630). At Puddletown a surveyor was appointed to supervise the construction of water meadows in

FIGURE 6.
A 'drowner' at work
at Charlton-All-Saints,
Wiltshire, in the 1920s.

1625, and the tenants agreed to contribute labour for the project, 'not sending boyes or woomen to the worke' (DRO D/PUD/B1/1/3). During the later seventeenth century John Snow, steward of Sir Joseph Ashe's estate at Downton on the Avon, south of Salisbury, not only supervised the large-scale scheme of water meadows there but was also employed by neighbouring gentry to survey and produce plans for diverting streams and creating meadows. On several occasions he was also sent by Sir Joseph Ashe to his estate at Wawne, in the East Riding of Yorkshire, to supervise and advise on the creation of water meadows there (Bettey 2003).

An indication of the difficulties and expense involved in implementing water meadow schemes can be seen in the accounts of Richard Osgood of Normanton, in the Avon valley near Amesbury, Wiltshire, in the period 1677–1681. A water meadow was already in existence there, but in 1680 Osgood was faced with the problem of renewing one of the main hatches in the strong current of the fast-flowing river Avon. He had several meetings with various relatives and others at the *Chopping Knife* in Amesbury, at which he sought their advice on how the work could best be done. Stone was ordered from the quarry at Chilmark, piles and oak posts were inserted in the river bed, masons were paid for building the 'land-fasts' or foundations for the hatch, working 'upp to the Arme Pitts in Water', and sawyers and carpenters were employed to cut posts of oak and elm to support the hatches. Evidently the strong current made the task particularly difficult, and the cost, including materials, labour and expenses at the *Chopping Knife*, came to more than £40 (WSRO 91/1).

Most schemes for water meadows involved only a single manor or group of tenants, but at Downton, mentioned above, a much larger project was successfully accomplished during the years 1665–90. This large manor, containing several villages and large farms, had been leased from the bishopric of Winchester in 1662 by the wealthy London merchant Sir Joseph Ashe. He thus secured control over a stretch of the Avon valley extending for nearly 5 miles (8 km) from Alderbury to the Hampshire border. This included the borough of Downton, for which Ashe became one of the two MPs. Ashe's steward at Downton, John Snow, inaugurated the complex and expensive scheme to create water meadows all along the valley of the Avon. The project involved careful surveying, major excavations to create subsidiary channels for the water, the construction of substantial hatches strong enough to withstand the current and to survive winter floods, and extensive work to prepare the surface of each meadow. Agreement had to be reached with mill-owners, fishermen and neighbouring landowners, and Snow displayed great skill and considerable patience in negotiation in order to overcome all the obstacles to the scheme, and pacify so many different interests. Not least of his problems was his employer. Sir Joseph had initially encouraged and supported the project but as it progressed he became increasingly alarmed and critical at the expense and bombarded Snow with complaints. Nonetheless, Snow was convinced of the major advantages which would follow from providing water meadows throughout his employer's estate, and he persevered in spite of all criticism until the whole project was successfully completed in 1690. By that time more than 250 acres (101 ha) of watered meadows had been created, at a cost which cannot have been short of £5,000. The scheme transformed the farming along this part of the Avon valley and more than doubled the value of the meadow land. It also provided an example which was quickly emulated on a smaller scale by neighbouring manors (Bettey 2003).

The widespread creation of water meadows along so many chalkland streams inevitably led to disputes with millers over the use of water. In 1646 the miller at Tarrant Rushton, Dorset, complained at the manorial court that his watercourse was blocked: 'obstructat et divertit causam innundationi pratori' (Salisbury MSS, Hatfield House, 9/3, Court Book of Tarrant Rushton 26 March 1646). Seventeenth-century leases of Waddock mill on the Piddle in Dorset all repeat the stipulation that the surrounding meadows could only be watered 'from each and every Saturday night unto each and every Monday morning and at all and every other Tyme or Tymes doing no prejudice unto the mill' (DRO D/FRA/T9). At Odstock, on the river Ebble near Downton, the construction of water meadows could only proceed after an agreement had been reached with the owner of Odstock Mill concerning the dates and times when the hatches providing water for the mill should be shut or drawn (WSRO 490/900). At nearby Nunton Lord Coleraine secured a lease in 1698 granting him £3 per annum for the use of the water from the stream which powered his paper mill (WSRO 490/894). Five farmers from Longparish, on the river Test in Hampshire, agreed to pay the miller there 2s 0d per acre per annum for the use of water from his mill leat on three days each week, and also undertook to have their corn ground at his mill (Bowie 1987, 157).

An agreement for a water meadow on the Itchen, north of Winchester, made between Robert Forder and Henry Sly in 1698, illustrates the complexity of the arrangements which were frequently necessary. The cost of building a hatch in the river to divert some of the water was to be borne equally by the two men; each was to pay 2s 6d per annum to the miller for use of the water; Robert Forder was to have the water for his meadow whenever he wished between 29 September and 25 December each year; Henry Sly could use the water from 25 December to 12 February each year. After 12 February there were further complicated rules for the use of water for 'two dayes and two nights in every weeke', and Robert Forder was to build a bridge over the tail drain from the meadow, 'sufficient to Carry over his Beasts or other Cattle' (HRO M94W/R18/1).

A lease for 7 acres (3 ha) of watered meadow beside a tributary of the Kennet in 1668 confirms John Aubrey's statement that water meadows were introduced in that district 'about 1646'. The document refers to the fact that the meadow was first watered or 'floated' in 1642, and the high rent which it commanded illustrates its usefulness in the farming economy: the annual rent of £11 for the 7 acres was more than double what could be charged for an unwatered meadow (WSRO 212A/27/27). In 1674 John Snow, the steward of Downton, attempted to justify his continuing to lay out large sums of money by producing the following comparison of values before and after the provision of water meadows:

> Brewers Farm 74 acres of meadow previously worth £74 per annum now worth £148.
> Witherington Farm 50 acres previously worth £20 per annum now worth £100 per annum
> New Court Farm 70 acres of meadow previously worth £80 per annum now worth £180 per annum (WSRO 490/904).

The value of the meadows

In 1808, 1,271 acres (514 ha) of water meadow on the Itchen to the south of Winchester were said to be worth £4,050, or £3 12s 0d per acre, whereas un-watered meadows were only worth a third as much. At nearby Compton, in 1758, unwatered meadows were let for 20s per acre, whereas watered meadows commanded 40s (Bowie 1987, 156–7). The extent and usefulness of watered meadows were recognised by various travellers and commentators. Early in the eighteenth century, Daniel Defoe was impressed by the farming of the downland and by the extent to which the increased size of sheep flocks had enabled farmers to extend their arable land on to parts of the downland:

> All which has been done by folding their sheep upon the plow'd lands, re-moving the fold every night to a fresh place, 'till the whole piece of ground has been folded on; this, and this alone, has made these lands, which in themselves are poor, and where, in some places, the earth is not above six inches above the solid chalk rock, able to bear as good wheat, as any of the richer lands in the vales … (Defoe 1927 edn, I, 284–5)

He went on to note that it was only by the dung of the sheep fold that these newly cultivated lands could be fertilised, since they were too far from the farmsteads in the valleys for dung to be carted to them (Defoe 1927 edn, I, 284–5). In a tour through the region in 1771, Arthur Young, the agricultural reporter and enthusiast for improved farming methods, commented on the sheep flocks, on the importance of the sheep fold in the downland farming and, above all, on the crucial role of the water meadows, without which such large flocks could not have been kept. He also provided an account of the water meadows throughout the year:

> They begin to water [after] the first autumnal rains; all they can throw over the land before Christmas they reckon the best, from the washing of new-dunged fields etc. They observe to lay it as thin under water as possible, so that the field retains its green colour; they leave it thus for three weeks or a month, and then draw it off, keeping the field dry for a month. After this they water it again several times during the rest of the winter. They begin

to feed them with sheep at Candlemas [2 February], and continue it till May-day; at that time they water for about a week or ten days, after which they are left for hay; the crop 1½ or 2 tons. Immediately on clearing the field the water is let on again for a week, which brings a growth for feeding, worth 10s an acre (Young 1771, pt. III, 284–5).

In view of the value and importance of water meadows to farming throughout the whole downland region of Wessex, it is remarkable that it was not until 1779 that the first full-scale account of the methods and benefits of watering was produced. This was George Boswell's *A Treatise on Watering Meadows* of 1779. George Boswell (1735–1815) was a farmer, land agent and enthusiast for improved farming methods and equipment, who lived at Puddletown. His book was dedicated to two local landowners, the Earl of Ilchester and James Frampton. Boswell provided a full and essentially practical guide which included detailed descriptions and diagrams of the layout of meadows, the construction of hatches and techniques of watering (Figure 7). His book achieved immediate success and a second edition was produced in 1790 (James and Bettey 1993).

By the end of the eighteenth century water meadows were at their fullest

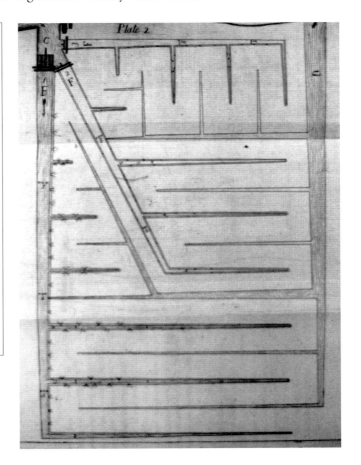

FIGURE 7.
Title page and
an illustration from
George Boswell's *A
Treatise on Watering
Meadows* of 1779.

extent and were of immense importance in the farming of the Wessex chalk-land. They were pushed to the limits: wherever the possibility of successful floating existed, water meadows were constructed. Their success was recognised in the tributes paid to their efficiency in the production of early grass and abundant, reliable hay crops by the authors of the *General Views* of each county in the region. Writing at a time of high prices and great prosperity for farming, all of these authors were enthusiastic about water meadows. John Claridge in 1793 claimed that 'the early vegetation produced by flooding is of such consequence to the Dorsetshire farmers that without it their present system of managing sheep would be about annihilated' (Claridge 1793, 34); while William Pearce, writing on Berkshire in 1794, urged that because the technique was so successful it should be extended 'to every tract of land that is capable of being watered' (Pearce 1794, 53). In 1794 Thomas Davis, the steward on the Longleat estate in Wiltshire, wrote:

> The water meadows of Wiltshire and the neighbouring counties are a branch of husbandry that can never be too highly recommended … None but those who have seen this kind of husbandry can form a just idea of the value of the fold of a flock of ewes and lambs, coming immediately with bellies full of young quick grass from a good water meadow, and particularly how much it will increase the quantity and quality of a crop of barley. (Davis 1794, 35–8)

The end of floating in Wessex

Water meadows retained their importance and their pride of place in the farming economy of the region throughout much of the nineteenth century, and remained as a vital element of some farms into the early twentieth century. After 1870, however, their importance slowly declined and they began to go out of use. The reasons for this decline are numerous: the influx of cheap grain from America, making homegrown wheat and barley much less profitable; the coming of the railways, which created a market for liquid milk for towns, so that milch cows replaced the folding sheep flocks as the principal stock on many farms; and the introduction of artificial fertilisers which removed the need for the sheep fold. New fodder crops and new strains of cultivated grass and clover provided a good alternative to meadow grass, and the water meadows were labour-intensive and expensive to maintain. In a few places, however, the traditional practices were retained and meadows continued to be watered. During the 1930s, the agricultural writer A.G. Street recalled life on his father's farm at Ditchampton just outside Wilton, before and after 1914. He stressed the central importance of the sheep flock in the farming system, 'whose continuous and regular folding over the land made the corn growing possible'. This was in spite of the fact that almost all of the profit made by the farm was derived from the dairy herd, whose milk was sent daily by rail to London. Although swedes, rape and kale were grown as winter feed for the flock, the water meadows remained crucial for providing grass during the 'hungry gap' of March and April. Much of the water meadow grass was reserved for the dairy herd since, with the low price of corn, 'the milk found the money which ran the whole farm' (Street 1932, 33, 36).

The final blow to the water meadows came with the agricultural depression of the 1930s, coupled with rising labour costs and the difficulty of using tractor-drawn machinery on the surface of the meadows. The now-neglected remains of many thousand acres of former water meadows can be found on either side of nearly all the chalk streams of Wessex. Abandoned weirs, hatches, channels and drains are still evident throughout the valleys. For three centuries these meadows were crucial to the specialised sheep-corn husbandry on which the farming of the downland region depended. Their introduction and rapid spread provided the first major victory in the long battle to improve agricultural productivity, and their role in sustaining the farming of the chalklands over such a long period was undoubtedly a triumph of English agriculture.

CHAPTER THREE

The Archaeology of Water Meadows

Christopher Taylor

Introduction

This chapter has three aims: to review the archaeological approach to water meadows over the last fifty years; to examine the present state of archaeology and water meadows; and to identify lines for future research. Until the 1950s archaeology, and especially analytical field archaeology, was largely concerned with prehistoric and Roman remains. Certainly medieval archaeology had begun, but it was still very limited in its scope; and post-medieval archaeology was virtually unknown. Water meadows thus attracted little attention: they were so recent, indeed sometimes still in operation, that they were not regarded as being of archaeological interest.

However, there was a larger problem. Archaeologists still worked on *sites*. Anything that would now be regarded as an historic landscape was largely unappreciated, despite the recognition of such landscapes even in the 1920s (Crawford 1928). The idea that irrigated water meadows, of recent date, stretching for long distances along river valleys, could be of archaeological value was almost beyond comprehension.

So, although archaeologists saw water meadows, they largely ignored them, perhaps because they thought they understood them. The Royal Commission on Historical Monuments, to its everlasting shame, never mentioned water meadows in its five Dorset volumes (RCHME 1952–75). One result of this academic short-sightedness was that many examples of water meadows were subsequently destroyed, or damaged beyond understanding, before they could be recorded. A survey in 1990 of Hampshire water meadows revealed that 40 per cent of what had originally existed had been destroyed, and that between 1970 and 1993 the condition of over one third of the water meadows in the county had deteriorated (Hampshire County Council 2002; 2003). In the years when the discovery and recording of all types of water meadows should have been undertaken, they were ignored.

The earliest studies of water meadows were by geographers, notably Carrier (1936) who wrote when the systems she described were still working. But it was not until the late 1960s and 1970s that water meadows as historic landscapes began to be appreciated. Work on the economic importance of water meadows also began at this time, initiated by Kerridge in Wiltshire (1953b; 1967). Soon afterwards historians and geographers began to record water meadows and to analyse their wider implications. Atwood (1964), Whitehead (1967) and Bettey (1977) all provided information and ideas that were new and relevant to the developing interest in water meadows. Archaeological work was much more restricted: witness the misguided and confused analysis of a catchwork system in Cambridgeshire (Taylor 1973, 175–7).

In the 1980s there began a more sustained phase of research on all forms of water meadows, which continues to this day. Much of this has involved the scientific background but there has been much new historical work (Bowie 1987; Cook and Williamson 1999; Bettey 1999). On the archaeological side there have also been major advances, mainly as a result of surveys on all types of water meadows. Examples include those by Wade Martin and Williamson (1994 and 1999b), Cowan (1982) and Cushion and Davison (2003, 190–3, 196–8) on bedworks; by Francis (1984), Jamieson (2001) and Riley and Wilson-North (2001, 128–9) on catchworks; and by Everson (1979) and Lillie (1998) on warping systems. Especially notable have been programmes of rapid survey such as those undertaken by Hampshire County Council (1999 and 2002) and the detailed ground surveys by English Heritage (McOmish *et al.* 2002, 132–6; NMR SP 00 NW 2).

Perhaps the most important result of this archaeological and historical research has been the realisation that irrigated water meadows are widespread. In their various forms they have been found, or are documented, in almost every part of Britain from Cornwall to the north of Scotland. Catchworks have been recorded all over Europe, from Scandinavia to Italy and the Balkans (Emanuelsson and Möller 1990, esp. 135–6; Rackham 1986, 338; Braudel 1992, 46). Another result of recent archaeological research is that it is now clear that the term 'irrigated meadows' covers at least five principal types, as well as a number of variants and hybrids.

Bedworks

The most familiar are the bedwork systems, much written about but, archaeologically at least, still not fully understood. Although bedworks appear to be the same wherever they are found, they vary in extent, form and arrangement. Variations include differences in the widths of the ridges, both within individual blocks and across whole layouts. These may be the result of differences in ownership or tenancies, of differences in original construction, or of subsequent changes. In small bedwork systems, the widths and heights of ridges seem to have been governed by the amount of water available. As early as 1653 better water flows were achieved by using progressively narrower and shallower carriers, and probably narrower ridges. Certainly, early nineteenth-century farmers in Dorset complained that very wide ridges were too flat to provide an adequate flow of water across the meadows (Whitehead 1967, 270).

The profiles of ridges can vary between high rounded forms, usual in Wessex, to flat-topped ones with sharply defined edges, as in Norfolk. The latter are difficult to explain, particularly as most of them display no obvious trace of carriers. Wade Martin and Williamson (1994) suggested that the lack of carriers at Castle Acre might have been the result of cattle trampling, although it is still curious that so little trace survives. Certainly, the flat-topped ridges there are original features of the system for similar ones were seen soon after their construction in 1804 at nearby West Tofts, where Arthur Young doubted their efficiency. The low gradients of the rivers and the high water table make it unlikely that the Norfolk bedworks were ever successfully worked in the Wessex way and it is even possible that a form of floating upwards was practised as well.

There is also considerable variety in the plan-form of ridges. Two recent detailed surveys in Wiltshire illustrate this (Figures 8 and 9; McOmish *et al.* 2002, 132–6). The system at Compton is highly geometric, with all the ridges being straight. That at Hindurrington, in contrast, has a much more irregular plan, the ridges there being markedly curved. There are also differences between the two sites in the design and construction of the hatches and sluices. At Hindurrington they are marked only by mounds of earth, while at Compton all are of brick. Both systems are, allegedly, of mid seventeenth-century date, but it is likely that much of what survives at Compton is considerably later in date than at Hindurrington.

The overall layout of bedworks also varies greatly. They range from a few

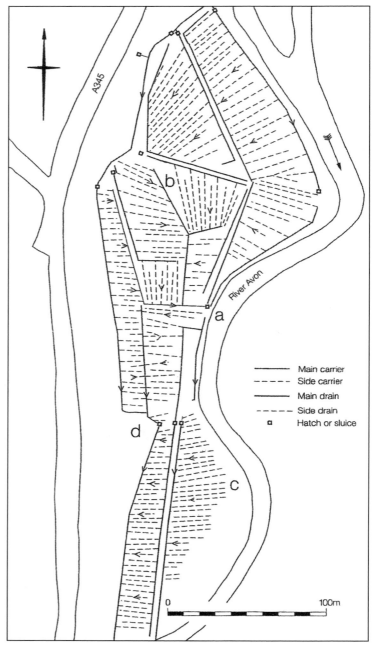

FIGURE 8.
Plan of water meadows at Compton, Wiltshire.
© ENGLISH HERITAGE

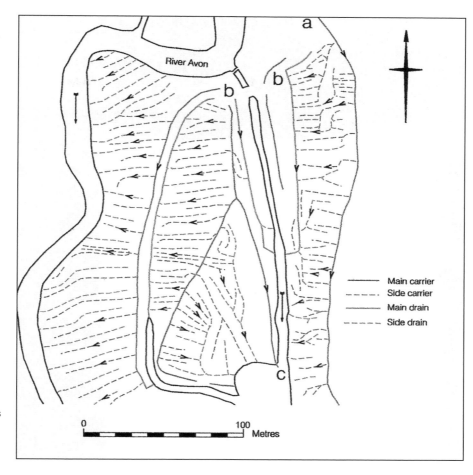

River Avon

a

b

b

Main carrier
Side carrier
Main drain
Side drain

c

0   100
                     Metres

FIGURE 9.
Plan of water meadows
at Hindurrington,
Wiltshire.
© ENGLISH HERITAGE

ridges contained within strip-shaped paddocks of between 1 and 4 acres (0.4–1.5 ha), as existed at West Harnham, Wiltshire, in 1787 (Steele and Tatton-Brown n.d., 1, 12, 13); through small rectangular blocks of up to 10 acres (4 ha), as at Britford, Wiltshire; to complicated layouts, extending for long distances along valleys. One of the most notable of the last is on the River Avon, south of Salisbury, created between 1665 and 1690. It includes two headmains, one just over 2 km long and the other 3 km (Bettey 2003, 164–5). In these cases, such differences in layout are evidently the result of differences in ownership. But, elsewhere, similar variations in the arrangement and size of blocks of bedworks known to have been in single occupancy or ownership at the time of their creation may have been the consequence of the process of construction (Cushion and Davison 2003, 196–8). A different interpretation of an irregular layout is possible at West Harnham. There, two small wedge-shaped blocks of bedworks, in the centre of the meadow, part of Cooper's Mead, may well indicate that the area in question is a later addition, squeezed into a gap left by the adjacent, more regular and earlier blocks of bedworks (Figure 10; Steele and Tatton-Brown n.d., 13, 14).

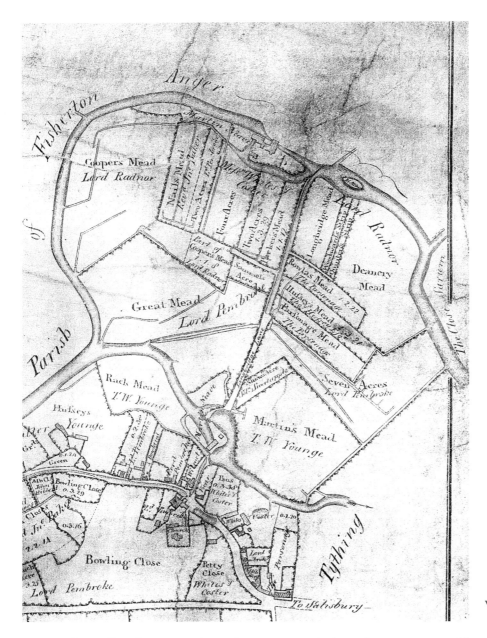

Irregular valley forms, resulting from geomorphological processes, could equally produce highly irregular arrangements of bedworks. The interleaved bedworks recorded at Stinsford, Dorset, have been interpreted as an adaptation to the differences in the gradient of individual meadows which may themselves have been the result of natural and earlier man-made processes (Whitehead 1967, 267–70). Likewise, some of the small blocks of bedworks on the Avon, south of Salisbury, seem to be related to the location and shape of areas of river gravel.

FIGURE 11.
An aqueduct on the water meadows at West Harnham in Wiltshire: a rare example of a mid nineteenth-century cast iron aqueduct, it takes a carrier moving water drawn from the river Nadder to a small meadow called 'Sammel's Acre' next to the Town Path between Salisbury and West Harnham.

FIGURE 12.
Nineteenth-century concrete cart bridge on the water meadows at Harnham.

27

One last explanation for the complex patterns of bedworks found in Wessex and elsewhere is that they may be the result of the 'prehistory' of the meadows. For these low-lying wetland areas had had centuries of use, and had been worked in a variety of ways, before the bedworks were constructed. The varying pattern of ownership and tenancies in the form of doles and allotments, as well as physical features such as drainage ditches, hedges and fences, may all have influenced the form of subsequent bedworks. Again, the meadows at West Harnham, Salisbury, illustrate this. The former long strips along the north side of the meadows, as well as other slightly curved examples on the east side of Long Bridge Lane, all of which existed as individual blocks of bedworks by the eighteenth century, could be the remains of medieval meadow allotments. If so, then the later bedworks were fitted into a pre-existing pattern of landholding. Support for these ideas appears on the enclosure map of West Harnham, dated 1787 (Figure 10), where one of the strips is called 'Parsonage Mead'. It has been suggested that this originally formed part of the lands of the Prebend of Coombe Bissett and Harnham. If so, this parcel of land must pre-date the construction of the bedworks. Further, in the fifteenth-century cartulary of St Nicholas's Hospital in Salisbury strips or 'swathes' of land are recorded in the south-east of the meadow (Steele and Tatton-Brown n.d., 21).

Another aspect of bedworks that recent research has elucidated relates to built structures. It has been possible to date bricks, to recognise mid nineteenth-century concrete and to identify the sources of stone used in the construction of sluices and hatches. The makers of nineteenth-century ironwork on hatches and winding gear have also been identified (Whitehead 1967, 272). Aqueducts of brick, cast iron and concrete, as well as inverted siphons, have also been recorded (Steele and Tatton-Brown n.d., 21–2) (Figures 11 and 12).

The final aspect to be noted in this examination of the present state of archaeological knowledge of bedworks is excavation. Little has yet been done, but the limited work at Twyford, Hampshire, has produced interesting results. This has included the examination of a bridge and the excavation of a section across a ridge, or pane. The latter showed that chalk had been brought in from outside the site for the construction (pers. comm. I.Wykes). Another excavation at Woodford, Wiltshire, has, similarly, shown that material was brought in from elsewhere in order to create the ridges (McKinley 2003, 9).

Catchworks

The second principal form of irrigated meadows are catchworks. At their simplest these consist of a headmain, or 'gutter', running along the side of a valley, from which the water was run down the slopes of the valley sides. The water sometimes passed through properly constructed hatches, as at Babraham, Cambridgeshire, but usually it was directed down the hillsides simply by digging temporary sections through the headmains. These systems are normally found on steeper valley sides, unlike bedworks, which are of necessity confined to flat-bottomed valleys. Being easy to construct and operate, catchworks are more widespread than bedworks (Carrier 1936, 119–20; Cutting and Cummings 1999, 159–60). Inevitably, a number of varieties have been recognised. An example of the simple type, consisting of only a headmain and hatches, was that (now destroyed) at Swaffham Bulbeck in Cambridgeshire (Taylor 1973, 175–7). More complex catchworks have subsidiary catch drains set on the hillside below the headmains, the better to distribute the water. Usually these drains, together with feeder trenches set at right angles to them, are carefully engineered, in order to allow an even spread of water across the slope. Many of the so-called gutter-ditch systems on Exmoor take this form (Jamieson 2001; Francis 1984). Sometimes, however, the catch drains were poorly constructed and levelled, with the result that the water was unevenly distributed. The seventeenth-century catchwork system at Babraham, Cambridgeshire, is an example of this (Taylor 2002, 114–15).

There are also differences in the location of catchworks, especially those in south-west England. Many can be called 'integrated systems', in the sense that the main supply of water passes through a farmyard, allowing the liquid manure produced there to be fed directly into the headmain and then washed down on to the meadows. Others, which can be termed 'detached systems', lie distant from their farmsteads and had no source of liquid manure, except when it was carted out to them (pers. comm. G. Brown).

A late form of catchwork, widespread in Britain from the mid eighteenth century, was sewage irrigation, in which urban sewage was either passed along drains to individual catchwork systems, as was done in Edinburgh from 1760 onwards, or carted to farms and mixed with the headmain water. By the nineteenth century such systems had become complex, with steam pumps sometimes being used to lift the sewage water into the headmains (at Romford, Essex, it is recorded that urban sewage was run on to bedworks) (Carrier 1936, 125–30; Mingay 1977, 170–1; Sheail 1996).

'Floating upwards' and warping

The third and simplest form of irrigated meadow is the 'floating upwards' system, whereby a stream is dammed and the ponded water allowed to flow back on to the adjacent meadowland. This method was probably a development of the natural flooding of low-lying ground, the benefits of which must have been obvious to early farmers, despite the danger of the development of anaerobic or even toxic conditions. The relative scarcity of this form in the archaeological record is probably not because it was rare, but because it is either very difficult to find or, perhaps more likely, because in southern England at least the original works have been overlain by later bedworks (Kerridge 1967, 163–4).

A possibly related type of floating upwards was noted by this writer in the 1990s on the North York Moors. In Sleightholmedale the Hodge Beck flows in deeply incised meanders cut through alluvial deposits. The river edges are bounded by low banks that appear to have been constructed to retain flood-water on the alluvial levels after the beck had returned to its normal course in the early spring. Another variant of floating upwards was described by Carrier (1936, 124–5), who described it as then being 'very modern'. Lines of underground pipes, fitted at intervals with vertical shafts, were laid across sloping meadows. Water was run into the pipes from ditches at the top of the meadow and the lower ends of the pipes were shut by 'sluices'. The water was forced up the shafts and flooded the meadow in an even way. This system was never adopted widely, perhaps because of the expense of both construction and management, but some examples may survive in redundant form.

A further variety of floating upwards, although so different in scale and purpose as to constitute a separate category, is 'warping'. This was the practice of running river water over adjacent land to ensure that the sediment was deposited and the soils enriched. The process is said to have been developed in the seventeenth century following drainage works that prevented natural flooding and led to soil impoverishment. Although recorded in the Fenlands of eastern England and in the Somerset Levels, the principle archaeological remains have been found alongside the River Trent in south Yorkshire and in Lincolnshire. There the short-lived eighteenth- and nineteenth-century warping drains are still visible as crop marks on aerial photographs, although buried by silt deposits. These deposits were laid down at a rate of between 100 cm and 400 cm a year and were 'refreshed' periodically by recutting the drains and reflooding. Although primarily for the production of rich pasture, warping was also used to improve arable land (Carrier 1936, 111–15; Everson 1979; Lillie 1998).

Another form of irrigation is visible in the eastern Fenlands. The unexpected fall in the surface of the southern peat fens from the mid seventeenth century onwards, following drainage works there, led to the necessity of confining the higher level rivers and main drains within artificial banks. To allow for the greatly increased flow in times of flood, these banks were set back from the watercourses they contained, thus forming wide areas of pasture known as 'washes'. These were regularly 'irrigated' during the winter months, providing valuable spring and summer grazing as well as hay. Whether 'washes' should be seen as natural or artificial irrigation systems is, however, debatable, and flood protection rather than irrigation was, and is, their primary function. Either way, their remains are a marked features of the Fens, the largest being the washlands created in the mid seventeenth century between the Old and New Bedford Rivers; these are 22 miles (35 km) in length and up to half a mile (0.8 km) wide. They were intended to hold, and still hold, the upland floodwaters of the River Ouse, now up to seven metres above the surrounding land, but from their beginnings they also provided summer pastures for cattle (Darby 1940, *passim*; Taylor 1999, 145).

In addition to these main types and their minor variants, it is also possible to recognise hybrid forms of irrigated meadows. One example has been discovered near the River Severn, at Buildwas in Shropshire. Work there has shown that the meadows were regularly flooded by natural inundation for several weeks early in the year. This, a form of floating upwards, was supplemented later in the season by artificial irrigation using a bedwork system (Brown 2002). A further hybrid form has been noted on the River Wey on the Hampshire/Surrey border. There, survey work has identified bedwork systems in the flat valley bottom that are combined with catchworks on the steep valley sides (pers. comm. I.Wykes). A further possible variant, already noted, are the curious Norfolk bedworks.

Despite a great deal of documentation relating to all the types of irrigated meadows, it has been very difficult to ascertain their beginnings. Warping, for example, is certainly documented in the seventeenth century but early references to the practice are not easy to distinguish from natural warping. Received wisdom is that the technique only became common after drainage work had taken place and cut off the natural supply of sediment (Kerridge 1967, 235; Williams 1970, 176) but, given the tradition of piecemeal fen drainage in medieval times, it is likely to have originated long before the seventeenth century (Darby 1940; Taylor 1973, 103–11).

The origins of bedworks have been more difficult to trace. Until recently it was believed that they were first developed in Herefordshire in the sixteenth century, probably by Rowland Vaughan. Now it is not so clear. Much of the surviving works on Vaughan's estate in the 'Golden Valley' appear to be of the catchwork variety or of hybrid form, and the remains of his own work seem to indicate that he practised a variety of 'floating upwards' (Cook *et al.* 2003, 157). Bettey (this volume) has shown that fully developed bedworks existed in Dorset by 1605 and he believes that this form of irrigation began in Wessex around that time. However, in reality their development probably took place over many years and is not clearly recorded in documents. Bedwork irrigation could have evolved from floating upwards, a technique that is so simple and so obvious a way of irrigation that it may well be medieval in origin. Certainly, floating upward seems to have been in use in Wiltshire by 1560 (Kerridge 1967, 254). Yet, as Cook *et al.* (2003, 158) have pointed out, John Fitzherbert described a method of irrigation very similar to bedworks as though it was an established practice as early as 1523, while field names seem to indicate the use of the technique at an even earlier date.

The catchwork method is certainly an old one. There is evidence for catchworks in Italy in the early twelfth century (Braudel 1992, 46) and Cook *et al.* (2003) have pulled together evidence to show that they probably existed in England in the twelfth century. There are, in addition, a number of archaeological sites of medieval date that appear to represent either catchworks, or else different and hitherto unknown forms of irrigation. An example of the first is at Rievaulx Abbey, Yorkshire (Cook *et al.* 2003, 160); an example of the second is at Bordesley Abbey, Worcestershire, where Aston (1972) has suggested that a group of narrow ditches, fed by a higher level channel, might have been a medieval irrigation system connected with the abbey.

Future research

The final section of this paper looks at what is now needed, in archaeological terms, in order to advance understanding of irrigated meadows. The first requirement is to establish how widespread the various forms are. It is clear that many sites of all types remain to be discovered. Places where documentary evidence indicates the former existence of irrigation systems should be examined first. One possible area to start might be Somerset. The Victoria County History for that county has references to a number of water meadow sites described in documentary sources from the fourteenth to the nineteenth century (Dunning 1974; 1978; 1985; 1992, *passim*). A rapid field check on these would prepare the ground for more detailed work.

The second archaeological requirement follows from the first. If water meadows are ever to be fully understood then large-scale, detailed ground surveys of as many sites as possible are required. Such surveys would help to elucidate some of the details of bedworks, such as the width, profiles and plan form of the ridges, as well as the arrangement of individual blocks. It should also be possible to establish relative chronologies and to find evidence of relaying. This latter feature is, perhaps, the most important, conceptually as much historically. There is a tendency to look at examples of bedworks, or catchworks, in places where irrigation schemes are known from historical sources to have been created in, say, 1650 and assume that what is on the ground today is all of that date. But in reality, the surviving archaeological remains are simply left by the last time the system was worked, not by the first. The centuries in between were usually ones of continuous alteration, deliberate or accidental; of improvement or decline. Through detailed surveys and documentary research it should be possible to establish the sequence of changes that took place. Certainly, some of the observable differences between irrigation systems must be the result of such alterations, involving on occasions complete relaying. Robert Wilson-North, who has worked on the gutter-ditch systems of Exmoor, recently wrote to the author: 'They seem very organic in that they are changed and altered regularly, depending on the needs and the resources of the farms' – a useful reminder that, with all forms of irrigation, what is being studied are dynamic not static systems.

Detailed surveys would also help understanding of historically or geographically important examples of surviving water meadows. A good example is the site at Dishley, Leicestershire. The bedworks there were created by Robert Bakewell, the pioneer eighteenth-century agriculturalist who was tenant here for many years. The irrigation system was associated with a canal that was used to float turnips from the fields to the farmstead (Kerridge 1967, 267; Wykes 2004, 39–40). Yet, despite an archaeological survey and field investigation, the exact form and method of working of this important system remains far from clear (NMR SK 52 SW 29). Likewise, although the bedworks created around 1676 for Lord Weymouth at Drayton Park in Staffordshire still exist, their depiction on Ordnance Survey maps is so inadequate that it is only possible to understand how they operated by examining the remains on the ground (Palliser 1976, 103).

Archaeological research should also be able to identify hitherto unrecognised forms of irrigated meadows for, inevitably, these will be found on the ground before they are discovered in documents. In Little Shelford in Cambridgeshire, for example, an island of meadow is bounded on the east by the River Cam and on the west by a medieval mill leat. Across the centre of this meadow is a degraded ditch, laid along what is the lowest part. This ditch is depicted on a map of 1748. Its function seems to have been to drain the meadow after flooding by water overflowing from both the river and the leat. Its date is quite unknown but it appears to be part of a crude form of irrigation.

Fieldwork could also identify traces of earlier systems of water management which might be confused with irrigation systems. One example is at North Meadow, Cricklade, in Wiltshire, where blocks of regularly spaced parallel shallow ditches are similar in plan to the surrounding bedworks. Only their unusual length and lack of ridging indicates a different origin. Although it has been suggested that they post-date the 1824 reallotment of the ancient hay lots, it is more likely that they relate to an earlier drainage system belonging to a very different grazing and hay-production regime (Whitehead 1981).

Archaeological excavation will also help to elucidate some of the minor details of irrigation systems, especially those of bedworks. It is doubtful whether there is scope for large-scale excavation and, given the essentially dynamic nature of these systems, much of the evidence recovered will probably relate to the latest phases of working. But details of hatches, sluices, siphons and bridges might help to better understand the mechanics of irrigation, as would further sections excavated across the 'panes', or ridges.

Another requirement is to establish more clearly when the various forms of irrigated meadows first developed. It may be that documentary research will produce evidence for the invention of bedworks, but this is probably unlikely. Ongoing archaeological survey work in Herefordshire, carried out in order to determine exactly what type of bedworks existed there and at what date, will certainly help. Establishing the origins of warping and floating upwards will probably be more difficult. Both methods must have been experimented with long before they are documented.

The medieval origins of catchworks are now established. But could they be earlier? The Romans were experts in water management, and what difference is there between an aqueduct and a headmain? There are numerous simple aqueducts known from Britain in the Roman period, from towns, the Welsh goldmines, at least one villa and from various military sites in the north (Burnham and Wacher 1990, *passim*; Johnson 1990, 70–8; Jones *et al.* 1960–2, 1–9; Richmond 1940–1). The best known is probably that at Poundbury, which supplied Roman Dorchester, Dorset, and is some 11 km long (RCHME 1970, 585–8). And what of the Car Dyke, which extends for 75 km along the western edge of the fens (Simmons 1979; Hall 1987, 28)? It has been interpreted as a canal, and more recently as a catchwater drain. But, equally, part of it could have operated as a headmain for catchworks of the type that existed in the nineteenth century on the eastern edge of the fens (Wade Martin and Williamson 1994, 26). There is also no doubt that Romano-British farmers could create ridges. They did so on their fen-edge farms in Cambridgeshire (Philips 1970, 124), while fieldwork on Salisbury Plain has revealed the remains of ponds and dams of Roman date (McOmish *et al.* 2002, 91, 104). None of this proves that there were Roman water meadows. It only suggests that they *could* have been created by a society in which farming organisation was far removed from subsistence agriculture, and in which surveyors and surveying instruments were widely available (Dilke 1971, 76–8). There is thus no reason why there should have not have been Roman water meadows, or even earlier ones. They will be difficult to find because most will have occupied the same areas as all of the later ones. But the main problem is in the minds of students of irrigated meadows. The possibility of Roman ones has not been considered, so they have not been looked for.

One last contribution that archaeology can make to the study of water meadows is during restoration work. When various types of irrigation systems are restored, detailed archaeological recording, advice and assistance is imperative. It is essential that structures are not damaged or carelessly altered and that all relevant information is collected and archived. This is best carried out by archaeologists (Hampshire County Council 2003).

So, at the end of this excursion through the past, present and future of archaeology and water meadows, what conclusion may be drawn? Much has been achieved, though not as much as should have been and there is still much to do. Archaeology, and in particular analytical field survey, has a great deal to offer in the elucidation of the subject. But it will only be successful if it is carefully integrated with historical and scientific research. Not for the first time has this writer advocated interdisciplinary work on subjects such as this (Taylor 1974). It seems to be a long time coming.

CHAPTER FOUR

'Floating' in Context:
Meadows in the Long Term

....

Tom Williamson

Introduction

The subject of 'floated' or irrigated meadows is, as the various other contributions to this volume make clear, one which fascinates scholars from a wide range of disciplines – history, archaeology and the natural sciences. But it is arguable that there are dangers in studying them alone, in isolation from the more general history of meadow land in England. The links between meadows, floated meadows, the exploitation of alternative sources of fodder, and the more general development of farming systems and landscapes are complex and intricate. Yet some attempt needs to be made to understand them, if we are to appreciate the true significance of 'floating' and the nature of its impact on English agricultural development.

Meadow and its alternatives

Meadows and pastures are sometimes confused in the popular mind and they were, in reality, often less distinct in the past then some modern commentators suggest. As the historian Bruce Campbell has noted in a medieval context, 'Meadow and pasture existed on a continuum and in practice the distinction between them is unlikely to have been as consistently drawn as the precise terminology of the documents might imply' (Campbell 2000, 78). Nevertheless, in theory the difference was clear enough. Pastures were directly grazed by livestock; meadows were only grazed for part of the year, for their main purpose was to supply a crop of hay, usually cut in June or July, which could be stored for use as fodder during the winter months. During the last century grass grew for an average of 250 days a year and, while in earlier periods warmer conditions might have extended the growing season in lowland England by at least 30 days, grazing would always have been in short supply by the late winter (Spedding and Diekmahns 1972, 19).

It is not surprising, then, that books on English or European agricultural history often describe meadows as 'essential' to medieval farmers, and they are invariably listed as a standard feature of the 'typical' medieval village: hay meadows were 'almost as vital to the community as the arable' (Bennet 1937, 43). In Eric Kerridge's eloquent phrase, they were 'the eyes of the land, like waterholes in the bush and artesian wells in the outback. No farmstead, no village could be built unless hard by them' (Kerridge 1973, 22–3). But, looked at in the long term, the importance of meadows should not be exaggerated. Hay was never the only source of winter fodder. Across much of northern Europe, including England, medieval and even post-medieval farmers made extensive use of 'leafy hay' – boughs cut from ash, oak, elm, holly and other trees, dried and stored for use as winter feed (Grieg 1988; Halstead 1998; Muir 2000; Fleming 1998) (Figure 13). Indeed, some trees were probably managed primarily with fodder in mind. Instead of being pollarded, they were *shred* – that is, the side branches were removed in order to stimulate epicormic growth: trees like fuzzy toilet brushes can still be seen in parts of western France. In medieval times, fodder crops could also be grown in the arable fields – beans were fed to cattle, oats to horses – while from the late seventeenth century the widespread cultivation of a new range of crops, most notably turnips, offered farmers a variety of important alternatives to hay.

It is important to remember the existence of these alternative sources of winter feed, not least because haymaking was a particularly laborious and complex task. When first cut, grass contains around 75 per cent water and to ensure good-quality hay this needs to be reduced to about 15 per cent before stacking, something achieved in part by the action of the wind, and in part by the heat of the sun. To encourage the escape of water the cut herbage has to be repeatedly turned over and shaken out, so as to expose as large a surface area as possible to the air. This was a very labour-intensive operation:

> The work must be carried out in such a manner that the herbage is not unduly exposed to the leaching action of rain nor yet to the bleaching action of the sun. Moreover, because the leaf of the grass plant has a higher feeding value than the stem, it is important to avoid over-drying which would cause the leaves to become brittle and liable to be broken off by the various implements used in the making and collection of the hay. (Robinson 1949, 286)

Stacking the hay while it contains too much sap can cause heating in the stack, leading to loss of digestible protein; stacking it when actually damp can lead to mould and rot. Hay-making with traditional tools thus required good weather, abundant labour, and the careful timing of farming operations: 'make hay while the sun shines', in the words of the well-known proverb (Figure 14).

Most medieval meadows, and certainly all the larger areas, were found on the floodplains of rivers and streams. There was a simple reason for this. Grass growth is most rapid between early May and mid June, after which hay was cut; but in central and eastern England, in particular, the size of the crop could be adversely affected by periods of low rainfall. Any deficiency in precipitation could, however, 'be offset by higher water tables which provide a continuously moist root range' (Robinson 1949, 256). Damp valley floors thus offered ideal conditions for hay production.

Hay meadows probably developed slowly across Europe. Indeed, the moist alluvial soils of the major valleys themselves often only came into existence in late prehistoric times, as deforestation of the surrounding catchments led to increased run-off, soil erosion and alluviation. When population levels were low, and where areas of pasture and, in particular, wood pasture remained extensive, there was less incentive to cut, store and dry hay, and meadows were thus probably of limited extent: the maintenance of large areas of meadow land 'implies a pastoral economy where the number of animals exceeds winter pasture' (Biddick 1989, 19–21). The wide range of herbs and shrubs found in wood pastures would have provided a longer grazing season than open pastures, the herbs and grasses of which offer little winter sustenance. Cattle, sheep and goats will browse off dry autumn foliage, even bare twigs; they will eat undershrubs such as brambles, which remain green well into the winter; while sheep in particular will avidly consume evergreens like ivy and mistletoe and cattle will eat holly with enthusiasm. Moreover, as already noted, the leaves of certain trees can themselves be cut and stored as 'leafy hay'.

In England, meadows are mentioned as early as the seventh century, in a law of King Ine of Wessex:

> If ceorls have common meadow or other deal-land to fence and some have fenced their deal, some never, and their common plough-acres or grass are eaten, go they then that own that gap and make amends to the others that have fenced their deal, for the damage that there be done. (Whitelock 1955, 368)

The importance of meadow land seems to have increased steadily as population rose in the course of the middle and later Saxon periods, however. The organic remains preserved in waterlogged deposits at Yarnton, on the Thames gravels outside Oxford, for example, show that the low-lying floodplain was managed as pasture in the sixth and seventh centuries, but from the eighth or ninth century areas were enclosed with ditches to exclude stock, and managed as meadow. Meadows frequently appear in Anglo-Saxon charters, most of which date from the period after the ninth century (Rackham 1986, 33). Nevertheless, as Rackham has pointed out, Domesday suggests that only around 1.2 per cent of the total land area of England consisted of meadow land at the end of the eleventh century. 'England in 1086 had at least as much arable land as in 1500, but much less meadow. Meadow had spread to every part of the country, but countless places only had an acre or two and many … managed without it' (Rackham 1986, 334–5). Rackham's precise figures need to be treated with a measure of caution, given how difficult and misleading a source Domesday can be, but it is hard to argue with his general point: that 'the practice of meadow was slow in being fully adopted' in early medieval England (Rackham 1986, 334).

In the course of the twelfth and thirteenth centuries the area occupied by

meadows was steadily extended, and by c. 1250 they occupied around 4 per cent of the land area of England (Rackham 1986, 337). Yet their continued scarcity, comparative to the needs of farmers, is reflected in their value. Acre for acre meadow might be worth two, three or even more times as much as arable. At Sherington in north Buckinghamshire, for example, the demesne meadows were leased in 1312 for six times the amount charged for the arable land (Chibnall 1965, 113). Moreover, the extent of meadows varied considerably from district to district. To judge from Bruce Campbell's exhaustive analyses of *Inquisitions Post Mortem* (IPMs), even in the early fourteenth century meadows remained in short supply 'in the east and west of England' (Figure 15):

In between lay the Midland plain. From Somerset and east Devon in the south-west to the Vale of Pickering in Yorkshire's north Riding in the north-east, it was in the clay vales of this broad diagonal band of country that meadowland was most consistently represented. Except on the wolds, few demesnes were without at least some meadow ... (Campbell 2000, 75–6)

FIGURE 13.
Ancient pollarded oaks at Staverton Park in Suffolk, probably the best surviving example of a wood pasture in East Anglia. In early times, branches would have been cut from shreds and pollards like these for use as browse in the autumn or as leafy hay in the winter.

FIGURE 14.
Hay-making at Lockinge, Oxfordshire in 1905.

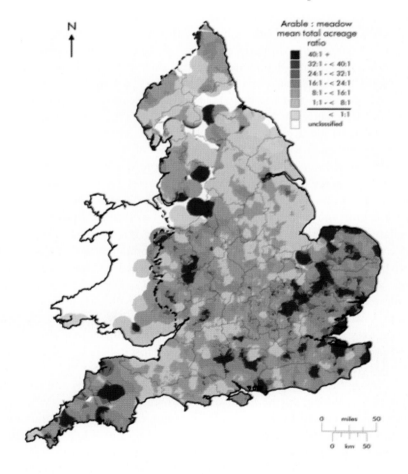

FIGURE 15.
The ratio of the arable acreage to the acreage of meadow in England in the early fourteenth century, as mapped by Bruce Campbell from the *Inquisitions Post Mortem*.

39

There were exceptions to this broad pattern: the Lea valley in Hertfordshire and Essex, for example – well outside the Midland Plain – had substantial areas of meadow. But for the most part comparatively small amounts of meadow appear on demesnes in the IPMs outside the Midland belt. Domesday seems to show a similar picture, although in detail this is distorted by inconsistent recording (Williamson 2003, 168–9).

These variations were in part a consequence of population pressure, and of the extent to which pastures and wood pastures remained extensive in each area. In part, however, they were the direct result of environmental factors, which ensured that in some areas it was harder to create meadows than in others. As Ault put it, 'In medieval times, meadows were the gift of nature, not the work of man' (Ault 1972, 25): and while this statement certainly underestimates the extent to which the relevant environments could be modified by human agency there were districts in which meadows could never be brought into existence on a large scale. Western Norfolk, for example, had little meadow recorded in either Domesday or the IPMs: here, as Campbell noted, 'with a low rainfall, sandy soils overlying chalk, and no substantial rivers, opportunities for the establishment of meadows were mostly non-existent' (Campbell 2000, 73–5). In the Chiltern Hills of Buckinghamshire, Hertfordshire and Oxfordshire, similarly, the floors of the principal rivers were narrow, their tributary streams few in number and their flow unreliable in the summer months. 'The absence of running water in the Chilterns away from the main valleys meant that meadows ... were often scarce, or of poor quality, or both, and frequently lay in small pockets' (Reed 1979, 100). Many manors in this district, to judge from thirteenth- and fourteenth-century extents, thus had negligible quantities of meadow. A survey of Westwick manor in St Albans (Hertfordshire) for example, made in 1306, describes 801 acres (324 ha) of arable, 44 acres (18 ha) of pasture and 32 acres (13 ha) of woodland, but only 11.5 acres (4.6 ha) of meadow, just over 1 per cent of the total area of the manor (Hunn 1994, 53). Further east, on the clay-covered dipslope of the East Anglian Heights, there was more meadow land but similar topographic factors ensured that it was still in limited supply compared with the situation in the Midlands. On the demesne at Little Maplestead in 1338 there was only a single acre of meadow for every 35 of arable; at Cressing at the same date the meadow:arable ratio was 1:36 (Larking and Kemble 1857, 168); while at Writtle in 1328 it was 1:33 (Newton 1960, 26–7). Meadow was, again, often scattered in small parcels. At Great Dunmow in Essex, for example, in c. 1250 the single acre of meadow land owned by the Knights Hospitallers lay in two separate plots (Gervers 1996, lxxxiv).

Elsewhere the development of meadows was hampered by the character of the deposits making up the floors of the principal river valleys. Where floodplains were occupied by alluvial soils, or some combination of alluvium and gravel, meadows often were, indeed, the 'gift of nature'. Only minor ditching and appropriate livestock management were required to produce good crops of hay. But much river valley land was too wet and waterlogged in its native state to provide the right conditions for growing good-quality grass and instead produced only coarse herbage, sedge and rushes, unpalatable to livestock and of low nutritional value. Here, *pace* Ault, the creation of meadows was not just a matter of management, of removing animals from the areas in question in the spring, in order to allow the hay crop to grow and mature. Instead it involved a positive act of investment and creation. The land needed to be drained by cutting ditches through and round it; and, in places, embanked against too frequent flooding from the main watercourse. Across large parts of East Anglia, in particular, potential meadow land was in very short supply. Rivers like the Waveney, the Stour, the Alde and the Gipping have wide floodplains but, instead of gravel and alluvium, these are largely occupied by areas of damp peat, or by mixtures of peat and clay, which give rise to the acid, peaty soils of the Mendham, Adventurers' 1 and 2, Altcar 2, Isleham 2 and Hanworth Associations (Hodge *et al.* 1984, 83–7, 90–92; 212–3, 231–5, 247–9). The minor tributary rivers are also, for the most part, dominated by peat, with only relatively limited pockets of alluvium. In their natural, undrained state the floodplains of most East Anglian rivers would have comprised rough fen vegetation.

With the appropriate investment such land certainly could be converted to meadow, and much already had been by the time of Domesday. Various references in thirteenth-century documents attest continuing attempts made to reclaim valley-bottom fens and upgrade them to hay meadows. At Forncett in Norfolk, for example, new areas of meadow land were still being created as late as c. 1300 (Davenport 1906, 31). Nevertheless, even in the early nineteenth century many valley floors in East Anglia were occupied by common fens, interrupted here and there by areas of private – usually in origin, demesne – meadows. Even these were often of poor quality, as Arthur Young noted in 1804:

> No person can have been to Norfolk without quickly perceiving, that in this branch of rural economy the county has very little to boast. Nowhere are meadows and pastures worse managed: in all parts of the county we see them over-run with all sorts of spontaneous rubbish, bushes, briars, rushes; the water stagnant … (Young 1804a, 370)

The abundance of meadow land in the medieval Midlands was itself the consequence of environmental and topographic factors. The great Midland rivers – the Nene, Ouse, Welland, Trent, Severn and their principal tributaries – ran, for the most part, parallel to the dips and scarps of the main geological formations. They meandered slowly across the vales eroded into the softer rocks, and had broad well-watered floodplains of the required character – they contained a variety of seasonally-waterlogged soils ideal for meadow land, principally those of the Fladbury 1 and Thames Associations – deep, clayey pelo-alluvial and pelo-calcareous alluvial gley soils respectively (Hodge *et al.* 1984, 194–6, 328–330). Most vills in the medieval Midlands thus had between 5 and 10 per cent of their land devoted to meadows (Fox 1984, 131).

At Laxton in Nottinghamshire, for example, there were 130 acres (53 ha) of arable, compared with 1307 acres (529 ha) of open-field land; while at Sherington in north Buckinghamshire the tenants held 858 acres (347 ha) in the two open fields, and 54 acres (22 ha) of meadow, and the demesne comprised 556 acres (225 ha) of land, of which 20 acres (8 ha) were meadow (Orwin and Orwin 1938, 103–4; Chibnall 1965, 92, 111–3).

These regional variations in the availability of meadow, especially between Midland and non-Midland areas, had a number of important implications for medieval agriculture and also, it can be argued, for the development of the landscape. In particular, the existence of large, continuous blocks of meadow land – and the need to mobilise labour at short notice to cut and turn the hay – may have been one factor in the development of the settlement pattern of tightly nucleated villages which was (and to some extent still is) characteristic of the Midland counties. Conversely, the more dispersed pattern of settlement found across much of East Anglia and the south-east of England, and in the west, may in part have arisen from the fact that meadows were more scattered, and in smaller ribbons, in these districts, or (in some cases) in very short supply: the extensive greens and commons found in many of these areas of 'ancient countryside' (*sensu* Rackham 1976), around which loose hamlets of farms and cottages often developed, may have been retained in order to compensate for the relative shortage of hay by providing additional grazing late into the autumn (Williamson 2003, 160–79). Many of these commons seem to have originated, not as areas of pasture, but as wood pastures.

The origins of floating

Historians have usually posited a fairly clear distinction between meadows on the one hand, and water meadows – managed by 'floating', or irrigation – on the other. In reality, the documentary sources suggest that there were a number of ways in which low-lying grassland could be managed in medieval and post-medieval times, including both periodic inundation, and forms of irrigation which fell short of full-blown 'floating' on the seventeenth-century Wessex model, as discussed, for example, by Bettey in this volume.

Fully developed floating of the kind practiced in seventeenth-century Wessex had three purposes: the forcing of an early growth of grass in the spring; the enhancement of the summer hay crop; and the improvement of the sward through dressing with lime and nutrients (and also through oxygenation, and the selection of particular species and the suppression of others, although the nature of these mechanisms was presumably unknown to contemporaries). It is important to note, however, that even in the nineteenth century not every irrigation system was intended to accomplish all of these aims. In upland areas of the north and west of Britain irrigation does not appear to have been much used to improve the summer hay crop. It was largely employed to encourage early grass growth and to improve the quality of the hill pastures more generally (Pusey 1849; Smith 1851, 140); as late as 1933 Heddle and Ogg advocated the use of irrigation by spring and stream water to improve yield and sward composition in the Scottish Highlands (Heddle and Ogg 1933). Soil quality was improved, causing artificially flushed areas to display better nitrification rates and a lower lime requirement than surrounding pasture, and higher potash and phosphate levels. In addition, farmers in these western areas seem to have been more interested than their counterparts in the east in the way that irrigation could be used to dress the sward with lime and manure, for the high rainfall in these regions steadily leached lime and nutrients from the pastures. Early agricultural writers make this emphasis clear: Seymour, writing about experiments in Carmarthenshire, advocated the dumping of 'mould, lime, and rotten dung' in the irrigation gutters (Seymour 1796, 460); while John Exter of Pilton, in Devon, suggested loading the irrigation water with 'the wash of adjoining roads, ditches, farm-yards &c.' (Exter 1798, 206). As Taylor has already noted, in many western districts farmers ensured that the main carriers passed through byres or farmyards before reaching the catchmeadows. Catchworks were particularly well suited to flush irrigation, and to the application of dissolved dung and lime, because the irrigation could be episodic – via discrete irrigation events – as well as taking the form of a steady-state system of flow applied over several days, the usual practice with bedwork systems. Despite the higher rainfall and even annual distribution of precipitation experienced in the hill regions of Britain (Cook 1999), 'flush irrigation' was well suited to the 'flashy' character of small upland catchments. Upland irrigation was commonly for two or three days at a time, repeated every ten to twelve days.

In marked contrast to the situation in the uplands of the north and west, in the east of England some water meadows were only irrigated to increase the summer hay crop. In Hertfordshire, for example, Arthur Young described in 1804 how the extensive water meadows on the river Gade in the southwest of the county – between Cassiobury and Watford, and around Uxbridge – were mown twice a year, but only the aftermath was actually fed (Young 1804b, 179). The meadows at Babraham in Cambridgeshire, probably laid out in the 1650s by the lord of the manor, Thomas Bennet, were similarly designed only to improve the hay crop, and do not seem to have been used for spring feeding (Wade Martins and Williamson 1994, 35). Irrigation began on Easter Monday or 'never sooner than two weeks before'. Here, moreover, 'No other art is exerted, but that merely of opening in the bank of the river small cuts for letting the water flow on to the meadows' (Young 1796, 177). Both Young, and that great propagandist of water meadows William Smith, expressed surprise at these features: 'circumstances … not easy to understand', in Young's words. Smith believed that 'The form of the works seems to prove, that they were not designed by any person from Wiltshire, and that the possessors are totally unacquainted with the management and utility of water meadows' (Smith 1806, 116–7).

Elsewhere there were other deviations from classic Wessex practice. Some early agricultural writers refer to the practice of 'floating upwards' – that is, simply flooding a meadow by controlling water at the point of exit. This was still practised in the eighteenth century in the Vale of Pewsey and certain Cotswold valleys: Marshall, in 1796, condemned this 'ancient method' on the grounds that covering the surface of the meadow with deep water for any length of time could kill off the grass (Marshall 1796c, 227–8): but, as a technique of enhancing the summer hay crop, this procedure may have produced certain benefits, if carried out with care and if, in particular, the flood was not allowed to stand for too long.

Most historians believe that true meadow irrigation, using a moving film of water, was first developed in the post-medieval period, some attributing it specifically to the Herefordshire landowner Rowland Vaughan, whose book *Most Approved and Long Experienced Water Workes* was published posthumously in 1610 (Vaughan 1610). Others place less emphasis on Vaughan, but nevertheless stress that the practice is first recorded in the chalklands of Wessex in the early seventeenth century, where it was employed largely in the form of valley-floor bedworks (Wade Martins 1995, 67–8). But meadow irrigation unquestionably has a much longer history, at least as a method of improving the quality and quantity of pastures, or of the summer hay crop, by dressing the sward with nutrients, increasing oxygenation or maintaining damp conditions through the spring and early summer months (Cook *et al.* 2003).

Folkingham, in 1610, thus advocated the practice of running 'water participating of a slimie and muddy substance' from 'land floods and fatte rivers' across meadow land in the spring (Folkingham 1610, 24); while as early as 1523 Fitzherbert advised:

> … yf there be any rynning water or lande flode that may be sette or brought to ronne ouer the medowes from the tyme that they be mowen vnto the begynning of May / and they will be moche bettr and it shall kylle / drowne / and driue away the moldywarpes / and fyll vp the lowe places with sande & make the grounde euyn and good to mowe. All maner of waters be good / so that they stande nat styll vpon the grounde. But especially that water that cometh out of a towne from eury mannes mydding or donghyll is best / and will make the medowes moost rankest. And fro the begynning of May tyll ye medowes be mowen and the hay goten in / the waters wolde be set by and ron another way. (Fitzherbert 1523, 58–9)

Fitzherbert appears to be describing an established, rather than an innovatory practice, and various references in local documents suggest that by the start of the sixteenth century irrigation was a normal form of management in some districts. In 1522, at Clent in Worcestershire, one Richard Sparry was indicted in the manorial court for building a dam in an area called Kings Meadow, which had led to the inundation of neighbouring land. Interestingly, he committed exactly the same offence in 1530. In both cases he was ordered to rectify matters, the court ruling that tenants could, by custom, divert water on to their own land for six days at a time between Pentecost (i.e., Whit Sunday, usually in May) and Michaelmas (28 September) (Amphlett 1890, 77–8; Currie 1998, 201). In parts of Devon, similarly, meadow irrigation appears to have been widespread by the sixteenth century: 'late sixteenth- and early seventeenth-century maps show systems that were both well developed and regulated' (Turner 2004, 29). We should also note the way in which various types of field name which, by the later seventeenth century, were used for irrigated meadows, are also recorded in documents pre-dating the publication of Vaughan's text, including *le Flote* in Kimbolton, Herefordshire, in 1610; and *Fludgatemedow* in Aston, Warwickshire, in 1490 (Field 1993, 90–1). The name 'Water Meadow' is itself frequently recorded in sixteenth-century documents. By the seventeenth century this term was usually used in its strict and correct sense, for a meadow intentionally irrigated with channels, sluices and so on, rather than in the wider modern sense of any area periodically overflowed by an adjacent watercourse without human assistance. When the term was used in the past to name a specific piece of land – highlighting, as it were, something different or distinctive about it – it seems particularly likely that it was being used in the correct sense (Field 1993, 91).

Indeed, field names indicating a degree of controlled flooding are found even in medieval documents. In 1339, for example, the field name *Le Flodgate-medewe* is recorded at Minshull Vernon, Cheshire. In his 1993 volume on field names John Field noted of this particular example that it was 'much earlier than the recognised starting date for managed water-meadows, and may relate to a precursor of the more elaborate technique …' (Field 1993, 91). But, in addition, a number of more explicit references exist to forms of medieval meadow irrigation. A list of the labour services due from the tenants of the Abbot of Westminster in a custumal from Pyrford, in the Wye valley in Surrey – an area noted in post-medieval times for its watered meadows – includes: 'damming the water, to overflow the Lord's Meadow ½d. Mowing the meadows for three days 3d' (Manning and Bray 1804, 154). Steward's accounts for Allerton Bywater in Yorkshire, from 1420/1, similarly record the payment of wages to one man for 'making several hatches (*gurgites*) in the flood banks (*ripe acquae*)', while payments were also made for 'flooding and raising the ditch near the king's highway for safe-keeping of the king's meadow there' (Moorhouse 1981, 697–8). All this should not in itself be surprising, given that various forms of meadow irrigation are now known from other areas of medieval Europe, including the Belgian Ardennes, Switzerland and (probably) Norway, and as far afield as Greenland (Mignot and De Meulemeester 2003; Arneborg 2003; Schmaedecke 2003).

Most of these early references, as already noted, probably relate to irrigation carried out in order to enhance the hay crop by dressing the sward with sediment and manure and maintaining the moisture levels in the soil during the early summer. Yet it remains possible that irrigation to force an 'early bite' was also being practised in some districts in medieval times, long before Vaughan wrote his famous work in the 1590s. The majority of early references probably relate to catchworks, or to floating upwards. But it again remains possible, as Taylor has already implied in this volume (above, p.31), that bedworks were already being used in some areas of England before the seventeenth century.

The geography of floating

As described in detail by Bettey in Chapter Two of this volume, in the course of the seventeenth century sophisticated forms of meadow irrigation – using, in particular, extensive systems of 'bedworks' – spread rapidly throughout the chalklands of Wiltshire, Dorset and Hampshire. Here floating was unquestionably used both to force an early growth of grass as well as to improve the summer hay crop. Many historians have rightly seen this as a major innovation, which had an impact not only on livestock numbers but also – because of consequent increases in manure production – on cereal yields. Indeed, Eric Kerridge saw floating as the key innovation of the seventeenth century, a period which, he argued, was of greater significance in English history than the more familiar 'agricultural revolution' of the eighteenth and early nineteenth centuries, in which livestock numbers, manure inputs and cereal yields were all increased through the cultivation of new fodder crops – turnips and clover – in regular rotations with cereals (Kerridge 1967).

Such was the scale on which the technique was adopted in the Wessex chalklands that by the middle of the eighteenth century there were few river valleys which were not being managed in this way. By the 1790s there were said to be between 15,000 and 20,000 acres (6,000–8,000 hectares) of watered meadow in south Wiltshire alone (Davis 1794, 34); while in Hampshire it has been suggested that 'during the eighteenth century, in particular, water meadows must have been pushed to the limits of areas where it was possible to construct them' (Moon and Green 1940, 377). In the course of the seventeenth and early eighteenth centuries floating also spread into the adjacent areas of Gloucestershire, and into parts of Surrey and Sussex, where the valley of the river Wye was noted for the practice. Irrigation – mainly using catchworks – was also, by this time, widespread in lowland parts of Devon, Somerset and the Marcher counties. In south Devon, for example, farmers were said to be 'every where sensible of the advantages of water meadows' by the end of the eighteenth century (Fraser 1794b, 40).

Yet away from these heartland areas in the south and west of the country these kinds of highly developed water meadows were only slowly and sparingly employed by farmers. Young thus asserted in 1804 that the practice of floating was 'Of very late standing in Norfolk: the experiments made are few, but they are interesting enough to promise a speedy extension'; while in Suffolk he noted that 'Of all the improvements wanted in this county, there is none so obvious, and of such importance, as watering meadows' (Young 1804a, 395; Young 1813b, 196). Similarly, in Kent in 1805 it was reported that 'the practice as yet has few friends' (Boys 1805, 164), while in the same year there were said to be no irrigated meadows at all in Lincolnshire (Young 1813a, 312–4). It is possible that relatively simple forms of floating, less complex and sophisticated than Wessex practice, were in fact more widespread in these districts than Young and other contemporaries imply (see Cummings and Cutting, this volume pp. 91–2) but irrigation was evidently a much less common and central practice in the eastern counties than it was in the south and west. There were, as we shall see (pp. 52–5, attempts to establish sophisticated water meadows in these eastern areas, especially in the late eighteenth and early nineteenth centuries, but they were limited in scale, and met with limited success.

Why was the technique embraced with such enthusiasm across much of southern and western England, yet apparently adopted more sparingly elsewhere? The answer seems to lie in a combination of economic, agrarian and environmental factors. Water meadows were, in general, most intensively developed, and most extensively adopted, in two main kinds of area. The first was in western England – in parts of Devon, Somerset, Herefordshire, Shropshire and to some extent Staffordshire. As a more nationally integrated economy had developed in the course of the fifteenth and sixteenth centuries these areas had come to specialise more and more in livestock production, but – as Campbell's map indicates (Figure 15) – many lacked extensive reserves of good-quality meadow land. A relatively warm Atlantic climate made the forcing of an early grass crop feasible on a regular basis, while in many places the character of the local topography allowed the construction of catchwork irrigation systems at relatively low cost. The adoption of meadow irrigation allowed larger numbers of stock to be kept through the winter months, and reflects the development in the course of the post-medieval period of a more highly capitalised and specialised regional farming economy in these 'pastoral vale lands' (Thirsk 1987).

The second, and more important, type of area in which floating was adopted was the 'sheep-corn' districts – areas of light land where specialised grain production depended on the inputs of manure provided by large flocks of sheep which were regularly folded on the arable land: but again, essentially those in which climatic conditions and topography were amenable to the technique, principally the chalklands of Wessex, for reasons explained by Cook in Chapter Eight of this volume. Larger sheep flocks meant that larger quantities of manure were available for the ploughlands, thus increasing yields, although in periods of slow population growth and low corn prices – between c. 1660 and 1750 – the profits to be made from wool and meat were probably of equal or even greater significance to farmers. Either way, the Wessex chalklands were an area in which meadow land was short supply, albeit this time in relation to the extent of the ploughlands. Improving the productivity of meadow land relieved a major bottleneck in the farming economy.

Outside these two broad regions – the Wessex chalklands, and the pastoral vale lands of the west – irrigated meadows were generally less desirable, more expensive to construct, less reliable and effective – or a combination of these things. In the Midland counties, for example, the relative abundance of meadow land discouraged the adoption of floating, while arable production in medieval and post-medieval times was much less tightly geared to the size of the sheep flocks. Much of the land in this region comprised heavy clays, giving rise to soils which were less easily leached of nutrients than the chalks and sands. Close folding was thus much less necessary than it was in light soil, sheep-corn areas, and was anyway often impossible for much of the year, for on these damp soils sheep were prone to foot-rot during the winter months, and could puddle and compact the soils (Fox 1984, 130–3). In many Midland townships folding was restricted to the late summer and early autumn (Kerridge 1992, 77–9).

In East Anglia and across much of eastern England, in contrast, it was not a lack of need so much as topographic and environmental constraints which discouraged the widespread creation of water meadows. Indeed, the same kinds of factors that had hindered the development of ordinary meadows also militated against their improvement through floating. Major rivers were often slow-moving and sluggish, their gradients gentle and their valleys wide, and with floors occupied by acid, peaty soils, inimical to successful drowning. Climate may also have been a factor. The sharp late frosts to which much of eastern England is prone caused damage to irrigated grass, while later in the year periodic droughts limited the opportunities for summer irrigation (Wade Martins and Williamson 1994, 35–7). As the agricultural writer Philip Pusey observed in 1848, irrigation was a technique poorly suited to the drier and colder parts of the country (1849, 465).

Floated meadows and the 'agricultural revolution'

There were, however, a number of important districts of sheep-corn husbandry which lay in the east of England: most notably, northern and western East Anglia, the Wolds of Yorkshire and Lincolnshire, the Chiltern Hills, and the North and South Downs. Intensification of farming in these areas in the course of the seventeenth and eighteenth centuries was hampered by the fact that production of grass and hay could not be significantly increased through floating, while in many of them meadows of any kind were in short supply. It is in this context that we should interpret the geography of the classic 'agricultural revolution' of the eighteenth and early nineteenth centuries.

If water meadows were the key form of agricultural improvement in the seventeenth century, the new rotations which combined cereal crops with regular courses of clover and turnips were the key development of the eighteenth (Chorley 1981; Sheil 1991; Trimmer 1969). Clover and other 'artificial grasses' provided high, nutritious yields of fodder, allowing more livestock to be kept. In addition, they had (as we now know) the useful characteristic of directly fixing atmospheric nitrogen in the soil. Turnips did not fix atmospheric nitrogen in this way, but they did provide an excellent source of winter fodder. The most famous form of 'improved' rotation, endlessly championed by propagandists like Arthur Young, was the so-called 'Norfolk four-course': a recurrent four-year cycle of wheat, turnips, barley and clover or other artificial grasses, such as sainfoin. In periods of sluggish population growth, when cereal prices were low – particularly in the period between c. 1660 and 1750 – these new crops allowed farmers to diversify into livestock (and especially cattle) production. When cereal prices rose once more after 1750, in contrast, they served to increase the supplies of manure, and thus produced higher crop yields. Clover, sainfoin (*Onobrychis sativa*) and nonesuch (*Medicago lupulina*) were being widely integrated into arable rotations in many parts of England by the end of the seventeenth century. The adoption of turnips, and thus of the 'improved' rotations, was more gradual, and more uneven. In part this was because where (as was often the case in sheep-corn areas) farming was organised on communal lines, in open-field systems, it was difficult to integrate the new crops into customary field courses.

There is little doubt that it was in parts of eastern England – where meadows had always been in short supply, and where irrigated meadows could not be successfully created to any significant extent – that the cultivation of the new crops, and especially turnips, first became widely established (Kerridge 1956; Wade Martins and Williamson 1999a, 99–130). As early as 1666 a lease for a farm in Horsham St Faiths in the heathlands of east Norfolk bound the tenant to leave twenty acres (8 ha) sown with the crop at the end of the tenancy, while a tithe account for Thorpe St Andrew in the same district, dating from 1706 (NRO MSS 16.023; NRO PD 228/51 (W)), shows that turnips already accounted for some 13 per cent of the cropped acreage of the parish. But it may have been in east Suffolk that the crop first came to be grown on a large scale. Certainly, in 1722 Daniel Defoe was able to describe how:

> This part of England is remarkable for being the first where the feeding and fattening of cattle, both sheep as well as black cattle, with turnips, was first practised in England, which is made a very great part of the improvement of their land to this day. (Defoe 1724, 1976, 86–7)

In the decades around 1700 turnips began to be grown in substantial numbers elsewhere in East Anglia. On the Houghton estate in north-west Norfolk they were being cultivated in small quantities as early as 1673, but the correspondence of John Wrott, agent for the Walpole family, suggests that they became more important as a crop around the end of the seventeenth century. By the 1720s and 30s their cultivation, together with that of clover, was being stipulated in leases for the tenant farms on the estate, often in large acreages – up to a fifth of the total area of the farm for each crop (Plumb 1952). On the nearby Raynham estate a series of letters dating from the period 1661–86 make no mention of turnips; but the next run of estate correspondence, from 1706, show that they were by then a well-established part of the farming regime (Saunders 1916). The surviving accounts from other estates in north-west Norfolk present a similar picture. On the Hunstanton estate's Downs Farm in Barret Ringstead, for example, the crop featured as part of the rotation in almost every field covered by a cropping account for the period 1715–19 (NRO LeStrange OA3).

The large-scale cultivation of turnips may have begun almost as early in the meadow-poor Chiltern Hills of Buckinghamshire and Hertfordshire. William Ellis, writing in 1736, believed that they were being grown in the district by the 1690s, and there are numerous references to both clover and turnips in probate inventories from the decades either side of 1700 (Richardson 1984, 254–5). But in other sheep-corn districts in the east of England the nature of open-field systems – more highly regulated and communal in character than those in the Chilterns or East Anglia – seems to have ensured that the uptake of the new crops was somewhat slower. Turnips could only be planted as a substitute for bare fallows where the right to graze across the fallow fields had been limited or abolished. Otherwise, any farmer planting his strips with turnips would simply have to watch them being eaten by the communal flocks. Turnips and clover *could* be cultivated in the open fields: but it required the agreement, or at least the tolerance, of the farming community as a whole. On the Wolds of Lincolnshire the cultivation of both crops is recorded as early as 1696 (Beastall 1978, 15): by 1754 it was standard on the Massingberd estates, often in regular, four-course rotations (Beastall 1978, 18). Yet some parishes in the district grew few if any of these crops even at the end of the century. In the Yorkshire Wolds, similarly, some farms were growing turnips as early as 1745, but Young was still able to describe their cultivation as a novelty in 1770. Nevertheless, their spread thereafter must have been rapid for Marshall described them in 1788 as 'the most solid basis of Wold husbandry' (Harris 1961, 65–6; Young 1770, vol. I, 181; Marshall 1788, 12).

While institutional forms – the existence of highly regulated open-field systems – might thus slow the spread of turnips in certain eastern sheep-corn districts, the contrast with the uptake of the 'new husbandry' on the chalk-lands of Wessex is nevertheless striking. In Wessex as a whole turnips were virtually unknown as a field crop before 1730, and rare before 1760 (Wordie 1984, 331–2). Indeed, as late as 1801, in Wiltshire parishes like Avebury, Tils-head, Berwick or Collingbourne Ducis, they still only occupied between 2 per cent and 4 per cent of the cropped acreage. In some parishes they were hardly grown at all, with only one acre out of 521 occupied by the crop at Durrington, for example (Turner 1982, III). As late as 1815 turnips were described as a 'comparatively late introduction' on the chalklands of Dorset, while in Hampshire even in 1813 they were said to be still 'gaining ground among the most respectable farmers' (Stevenson 1815, 251; Vancouver 1813, 175). Even then they were not necessarily used as they were in the east: 'In the open fields, turnips are often sown in the fallow season, merely for the purpose of plough-ing them in, before wheat is sown on the same land' (Stevenson 1815, 263).

There can be little doubt that the divergent patterns of agrarian development in eastern and western sheep-corn districts in the course of the eighteenth and early nineteenth centuries owed much to variations in the extent of meadow land – and to variations in the extent to which the productivity of meadows could be improved by irrigation. Kerridge's arguments for an 'agricultural revolution' in the seventeenth century, based in large measure on the floating of the meadows, might have been more convincing if they had been limited to the chalklands of Wessex: conversely, the more familiar 'revolution' of the eighteenth century was clearly pioneered in those arable, and especially sheep-corn, districts of eastern England where floating could not be adopted on a large scale and where, in many cases, meadows of any kind had always been in short supply, of poor quality, or both.

Conclusion

Floated meadows should not be seen as something clearly distinct from unwatered meadows, and some forms of management fell uncomfortably between these two neat categories. The sophisticated bedworks which transformed agrarian production on the southern chalklands in the seventeenth and early eighteenth centuries had probably developed gradually, out of simpler forms of irrigation which had perhaps had the more limited aim of enhancing the summer hay crop and improving the quality of the sward: indeed, irrigation mainly or entirely for these purposes continued to be practised in many districts of England throughout the post-medieval period. Either way, meadows in their various forms have played a key role in the history of English agriculture, and variations both in the extent of meadow land, and in the extent to which its productivity could be improved through irrigation, have been major factors over the centuries in generating regional differences in agricultural practices, and perhaps in the organisation of the rural landscape. The history of 'floating', in other words, should be considered as one strand in a more complex story: only by placing the practice within this wider historical and environmental context can we appreciate not only its true importance in England's agricultural history, but also the impact of its *absence* from particular localities.

CHAPTER FIVE

The Later History
of Water Meadows

··

Hadrian Cook and Tom Williamson

Introduction

Water meadows are usually discussed as an essentially early modern form of agricultural improvement and, as we have seen, they formed a major element in Kerridge's proposed 'agricultural revolution' of the seventeenth century (Kerridge 1967, 251–67). But it is arguable that it was in the period after c. 1780 that water meadows achieved their most sophisticated and complex forms, and certainly their most extensive geographical distribution. Rather than being relics of an early modern agriculture, soon to be succeeded by different methods of improvement, a large number of surviving systems were, in fact, the creation of the agricultural revolution and High Farming periods of the late eighteenth and nineteenth centuries. Indeed, nineteenth-century observers seem to have regarded meadow irrigation as an essentially modern, rather than as an established and old-fashioned, technique. In George Eliot's *Mill on the Floss* the miller, opposing (in the traditional manner of millers) the extension of the irrigation scheme of his neighbour, symbolises conservatism, while his opponent represents the forces of modernity.

The spread of floating

The diffusion of water meadows from their southern and western heartlands is one notable feature of the period. It is signalled by the appearance of a number of key texts devoted to the subject, most notably George Boswell's *A Treatise on Watering Meadows* of 1779, Thomas Wright's *An Account of the Advantages of Watering Meadows by Art* of 1789, and William Smith's *Observations on the Utility, Form and Management of Water Meadows* of 1806 (Boswell 1779; Wright 1789; Smith 1806) (Figure 16). All combined sound practical advice with a strong polemic tone, the latter being shared by the writers of the *General Views* on the agriculture of the various counties, published in the decades either side of 1800. Keen advocates of agricultural improvement were clearly puzzled by the failure of water meadows to spread far beyond the south and west of England, Parkinson declaring of Huntingdon in 1813:

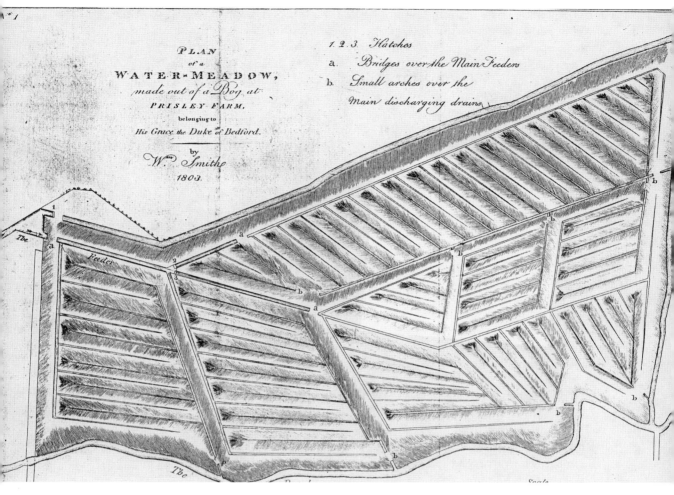

PLAN
of a
WATER-MEADOW,
made out of a Bog at
PRISLEY FARM,
belonging to
His Grace the Duke of Bedford.
by
Wᵐ Smith
1803.

1. 2. 3. Hatches
a. Bridges over the Main Feeders
b. Small arches over the
Main discharging drains

What a pity that some nobleman or gentleman would not cause this improvement to be adopted in a few parishes, and then the example would cause farmers and others to adopt the same plan in similar situations, for their own advantage. (Parkinson 1813, 313)

Some attempts to create watered meadows had, in fact, been made in the eastern counties during the eighteenth and possibly the seventeenth century. An undated map of Sedgeford in west Norfolk, for example – which certainly pre-dates the enclosure of that parish in 1766 – shows what is clearly a small floated meadow on the Heacham river (NRO LeStrange OB3). There are signs, too, that primitive forms of 'drowning', involving the simple controlled inundation of riverside meadows, may have been more widely practised by local farmers, at least in certain areas, than agricultural writers of the eighteenth century suggest (see pp. 91–2). But there is little doubt that, across much of eastern England, properly engineered and managed water meadows remained small in scale and few in number. Indeed, Young, writing in 1796, thought that the example at Babraham in Cambridgeshire was 'the only watered meadow of any consequence on this side of the kingdom'; while William Smith in 1806 was confident that until very recently floating had been unknown in the eastern counties (Young 1796, 177; Smith 1806, 1–12). Indeed, it was only in the period after 1790 that the water meadows of classic form were adopted with any real enthusiasm in the east, as agricultural prices rose spectacularly during the Napoleonic Wars.

The best evidence comes from Norfolk, probably because it was here that the technique was most widely adopted (Wade Martins and Williamson 1994). A number of new water meadows were created in the decades around 1800, especially in the valleys of the rivers Nar and Stiffkey, and in other light soil, sheep-corn districts in the west of the county. Young, in 1804, described the meadows which had been recently installed at Riddlesworth, Lynford and West Tofts, all in Breckland, and at Wighton and Houghton St Giles in the valley of the river Stiffkey (Young 1804a, 396–8). He noted that, at the time of writing, others were being planned at Billingford and Heacham: the latter at least was certainly constructed, for the earthworks – in the form of bedworks covering around four hectares – still survive. Further systems soon followed, including those in the Nar valley at West Lexham, laid out by William Smith himself around 1804–5 and described in some detail in his treatise of 1806 (Smith 1806); the even larger and more elaborate system just downstream at Castle Acre, laid out shortly after 1808 (Figure 17); and, further up the valley and a little later, systems at Kempstone, East Lexham and Mileham (Wade Martins and Williamson 1994, 23–7). Other examples were created in the early nineteenth century – to judge from comments in Smith's *Observations* – at Taverham (6 miles (9.7 km) north-west of Norwich), Beechamwell, in the Swaffham area, and at Waterden and Easton.

The impetus for this expansion of floating eastwards was in large measure financial – the high grain prices of the Napoleonic War years. But there are also signs that water meadows were something of a fad, a fashionable improvement adopted by large landowners and gentleman farmers with a particular interest in agricultural innovation. It is noticeable, for example, that a high proportion of the Norfolk examples were created on Holkham estate farms: indeed, the estate's owner, the noted agriculturalist Thomas William Coke, began in 1803 to offer through the Norfolk Agricultural Society an annual prize of 'A piece of plate to the value of 5 guineas to such person as shall convert the greatest area of waste or unimproved meadows into water meadows in the most complete manner' (*Annals of Agriculture* 1803, 322). Nevertheless, and *pace* earlier comments (Wade Martins and Williamson 1994 and 1999b), it is now clear that irrigation was not *only* adopted on the estates of improving landlords and by large tenant farmers. The archaeological evidence indicates that smaller landowners, people below the notice of the agricultural writers and whose activities have left less copious documentation, were also experimenting, albeit on a small scale. Recent work by the National Mapping Programme has located a number of new examples in the east of the county, at Lessingham,

Roughton and East Ruston, which have been destroyed since 1946 (Albone, Massey and Tremlett 2004, 551). One example survives in earthwork form at Dilham: a bedwork system, it is fed from the river Ant Navigation, and must therefore post-date the river's construction in 1825.

In the adjoining counties the pattern of adoption appears to have been much the same. Little interest was shown in watering, at least in its developed forms, before the late eighteenth century, but the high prices of the Napoleonic War years encouraged limited adoption, largely by major landowners and large tenant farmers. In Lincolnshire, for example, the single example of irrigation recorded in 1813, at Osbornby, was said to be the work of 'that

In Scotland the story was similar. Simple catchmeadows may have existed in lowland Scotland from an early date: in the 1790s James Robertson commented that 'From the remains of old tracts, made for conveying water, it appears that the practice of watering land, in various parts of Stathern, was more frequent in former than in later times' (Robertson 1794, 29). But, as Iain Fraser comments in a recent review of the evidence, 'The years between 1798 and 1810 were to see a wave of interest in the subject', with the technique being strongly advocated by writers such as J. Smith (1799). This was followed by a second period of 'renewed popularity from the mid 1820s to about 1850' (Fraser 2001, 133). Several extensive and elaborate systems were created at the start of the century, such as those on the Duke of Buccleugh's estates at Mosspeeble, while George Stephens listed a number of places where irrigation schemes had recently been instigated in his book *The Practical Irrigator. Being an account of the utility, formation and management of irrigated meadows, with a particular account of the success of irrigation in Scotland* (Stephens 1834). These included Lord Willoughby de Eresby's farm at Forr, to the south-west of Crieff; the Duke of Athol's estate at Dunkeld; the Earl of Mansfield's estate at Scone; Strathallan in Strathearn; and Glendevon in Tayside (Stephens 1834, 52–7, 63–72). Many examples were in the lowlands of eastern Scotland, and some of these took the form of bedwork systems, closely modelled on those in southern England, but catchworks were also widely created, especially in upland areas. The benefits of irrigation on one particular sheep farm were described in some detail in an article in *JRASE* in 1846 (Roal 1846), and other examples are noted in the *New Statistical Account* at Inchbrakie in Tayside (in 1838) and at Cortachy and Clova in Angus (in 1842) (New Stat Acc x, 509; xi, 449). As in eastern England, however, uptake remained uneven and most water meadows were abandoned in the late nineteenth century:

> In the first half of the 19th century the water meadow was seen as an outstanding example of the application of scientific principles to agricultural improvement, and a necessary component of any improving estate. Later experience indicated that the technique was not wholly appropriate to the climate, geology and farming practices of the north, and this left the meadows vulnerable to the vagaries of economic change and further agricultural development (Fraser 2001, 143).

Here, once again, floating was essentially a technique adopted by large proprietors: water meadows were created by 'major landowners rather than tenant farmers' (Fraser 2001, 143).

In Wales, similarly, water meadows experienced a period of popularity from the end of the eighteenth century. Clarke, in 1795, proposed using springs and brooks on hillsides in Brecknorshire as much as 'twelve hundred feet above the beds of the main rivers' during both winter and summer, while Seymour, writing in 1796, described the results of experiments in hillside irrigation in Carmarthenshire, in which a rivulet was dammed and water led along the hillsides in parallel gutters (Clarke 1795; Seymour 1796, 460). He was not alone; two years earlier it had been reported that:

> Many occupiers of the Lands, whose situation admits making Watered meadows, are becoming attentive to that mode of improvement; and the increase of Watered Meadow Lands is conspicuous of years throughout this district ... However, there yet remains a great deal to be done. (Hassall 1794, 12)

While in neighbouring Pembrokeshire:

> The benefit of watering meadow lands begins to be generally acknowledged in the County; and the opportunities for applying this valuable branch of improvement are almost every where to be met with. The great impediment is, the want of skill in conducting the water, and in making ponds. (Hassall 1794, 11)

In lowland districts in the north and west, especially in Scotland, water meadows seem to have been used in a similar way to those in southern England, both to provide an 'early bite' and to enhance the summer hay crop. Even in upland areas, the benefits brought by the early warming of the grass were appreciated. The average water temperature for springs on Exmoor during 1854 and 1855 was 8 °C (Smith 1856a), even when air temperatures were below zero. Measurements taken in April 1998 of the water from a spring at Wydon Farm in the Brendon Hills similarly showed a temperature of 8.5 °C, compared with an air temperature of 5 °C: the temperature of the water overtopping the carrier and irrigating the sward was at 8 °C, and even that in the tail drain was 7 °C. But while such upland catchmeadows were used to force an early growth of grass, they were not much valued as a way of increasing hay production (Pusey 1849, 471–4); and perhaps of most importance to moorland farmers were the benefits of 'flush' irrigation, the dressing of lime which the water brought to the pasture, and the way that irrigation could be used to treat the sward with suspended nutrients, especially dung from farmyards. Robert Smith, the Knights' steward on Exmoor, was a particular proponent of 'washing-in', proposing that irrigation should be seen at least in part as a way of transferring farmyard dung to meadows (Smith 1851, 143). Many of the nineteenth-century catchworks created in the uplands had a main carrier which first passed through the farmyard or cattle sheds before being used for irrigation. At Wydon Farm, near Minehead in Somerset, for example, a spring issues above the farmyard, runs through a redundant drinking trough and beneath a cattle shed from where it re-emerges into a pond. A large plug in the bottom of the pond released the contents into a culvert which still runs beneath the road, and thus on to the meadow (automated versions of such a system were described by some writers (e.g., Smith 1851)). On Exmoor, the farmsteads built by the Knight family for tenant farmers were all provided with gutters which carried dung and urine, mixed with water, on to the catchworks.

As already noted, catchworks were well suited to flush irrigation, and were a particularly good way of applying suspended particles of dung and dissolved lime because irrigation could take place as a series of discrete episodes: continuous flow over long periods, of the kind usual in bedworks, was not required. Catchworks were also cheap to construct. True, many nineteenth-century writers advocated thorough preparation of the ground before the gutters were installed, including paring and burning, and ploughing and harrowing, in order to break up soil pans to facilitate drainage. But even so, in the mid nineteenth century catchmeadows probably cost, in most upland locations, between £6 and £11 per acre (Smith 1856b), while the cheaper method advocated by Bickford (see below) was claimed to cost less than £5 per acre – at a time when bedworks cost between £30 and £50 an acre (Northcote 1855, 115). Such construction costs would be attractive to hill farmers with little capital. So too would the fact that individual catchmeadows were normally operated on an *ad hoc* basis by individual farmers – there is little evidence here for the kind of professional 'drowners' found in Wessex.

Numerous articles in the *Journal of the Royal Agricultural Society of England* in the middle decades of the nineteenth century described particular schemes in the north and west, and especially in upland areas, and advocated their further extension, as did articles like Barker's 'On water meadows as suitable for Wales and other mountain districts' of 1859 (Barker 1858). Noted improvers like Philip Pusey waxed lyrical on the subject: water-meadows were the 'talisman by which a mantle of luxuriant verdure might be spread across the mountains and moors of Wales and Scotland, of Kerry and Connemara' (Pusey 1845, 521). But many areas in the north and west of Britain seem to have completely shunned the technique, such as Lancashire (Garnett 1849), and even in a county like Cornwall uptake seems to have been very limited (Fraser 1794b). Catchmeadows do not seem to have caught on in Scotland to anything like the extent that writers like Stephens advocated (Stephens 1834), while Read noted in 1849 that irrigation in South Wales remained limited and unsophisticated in character:

> Not practised further than turning the water from the yards or a road-side ditch across a meadow. This is done simply by running a furrow out with a plough, and when that portion of the land has received its share another is drawn in a different direction and the former one is closed up (Read 1849, 139).

On balance it seems that with the exception of certain areas like Exmoor and the Quantocks, and the home farms of large aristocratic estates, irrigated meadows remained relatively rare in upland regions, and in the north and west of Britain more generally. As *Stephens' Book of the Farm* put it in 1908, although 'Great benefits may also at times be derived from water-meadows in some of our Highland districts', it remained the case that 'the northern climate is so different, in rainfall, temperatures and variableness, that the good effects of artificial watering cannot be relied on to the same extent as in the south' (MacDonald 1908, 355).

Improvement and innovation

The late eighteenth and nineteenth centuries thus saw the practice of floating expand into new areas, albeit with mixed success. The same period also saw the development of more complex and sophisticated systems of irrigation, the use of new materials, and the development of new methods of construction and management. There were, for example, experiments with simpler, and cheaper, systems. Permanent catchwork gutters required large volumes of water to fill them and were not suitable where only small springs and streams existed. Barker (1858) described a method for cutting parallel gutters 9 inches (23 cm) wide, 6 inches (15 cm) deep and about 4 poles (20 m) long where water was in relatively short supply. This system of 'Bickfordising' (after a Mr Bickford of Crediton) was advocated by a number of nineteenth-century writers, including Northcote:

> A carriage gutter is in the first place cut in the usual manner along the line of the highest ground. The inclination of this gutter should, if possible, be such as to give a fall of 2 inches in 4 land yards, or 1 in 396; but if such a fall cannot be obtained, a much less rapid one will answer the purpose. The gutter should be about 1 foot [30 cm] wide, and about 6 [15 cm] inches deep. These dimensions, which are much less than those usually observed upon the old system, should be gradually contracted as the gutter approaches its termination, so it may die into the ground. Below this carriage gutter, and in the same direction with it, are cut other and smaller gutters, which are perfectly level from end to end, and which successively catch the water as it trickles over the land immediately above them, collect it, and when they are full, begin to overflow at all points at once, and distribute it evenly to the next section. (Northcote 1855, 114)

A 'gutter plough' could be used to produce horizontal gutters, some 4 inches (10 cm) wide and 3 or 4 inches (7.5–8 cm) deep, which could be recut each year. The plough was an adaptation of the horse hoe, with two upright knives in front (which cut the sides of the gutter) and a furrow slicer behind which severed the furrow slice from the ground.

Nevertheless, while simpler methods of construction were explored in this period, in archaeological terms the most striking feature of late eighteenth- and nineteenth-century irrigation systems is their flamboyant sophistication. Particularly spectacular were some of the new meadows created outside the traditional homelands of watering, in the Midlands or east of England, such as the extensive and complex systems created for the Duke of Portland at Mansfield and Ollerton in Sherwood Forest in the 1820s, described in some detail by John Dennison in 1840. These were mainly catchworks, covering in all some 120 hectares, which were fed by more than 11 km of carriers, lined with limestone and lime to prevent erosion (Figure 18). The difference in levels between the 'flood dyke' (the main carrier) and the river was approximately 17.6 m, the slopes of the hillsides generally being in the order of 7 degrees.

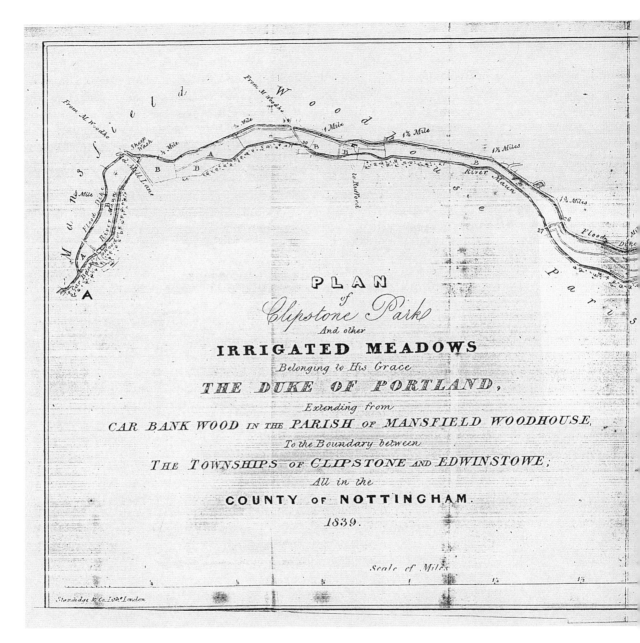

PLAN
of
Clipstone Park
And other
IRRIGATED MEADOWS
Belonging to His Grace
THE DUKE OF PORTLAND,
Extending from
CAR BANK WOOD IN THE **PARISH** OF **MANSFIELD WOODHOUSE,**
To the Boundary between
THE TOWNSHIPS OF **CLIPSTONE** AND **EDWINSTOWE:**
All in the
COUNTY OF **NOTTINGHAM.**
1839.

Scale of Miles

Dennison describes how the land, previously rough grazing of little value, had been cleared of gorse and heather, levelled and ploughed before the system was laid out. The enterprise cost no less than £38,000, a vast sum, but Dennison believed that the investment was worthwhile because of the benefits the meadows brought to the *arable* land of the estate:

> They afford, through the cattle fed in yards, such a weight of manure for other land, that large districts have, by these means, been brought into

FIGURE 18.
Plan of the water meadows at Clipstone Park, Nottinghamshire, as illustrated by John Dennison in 1840.

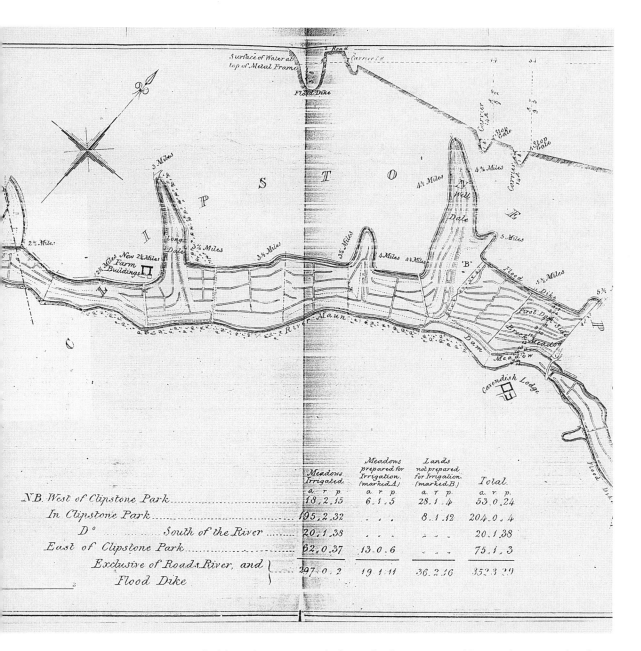

	Meadows Irrigated.	Meadows prepared for Irrigation. (marked A)	Lands not prepared for Irrigation. (marked B)	Total.
	a. r. p.	a. r. p.	a. r. p.	a. r. p.
N.B. West of Clipstone Park	18.2.15	6.1.5	28.1.4	53.0.24
In Clipstone Park	195.2.32	. . .	8.1.12	204.0.4
D° South of the River	20.1.38	. . .	. . .	20.1.38
East of Clipstone Park	62.0.37	13.0.6	. . .	75.1.3
Exclusive of Roads, River, and Flood Dike	297.0.2	19.1.11	36.2.16	352.3.29

profitable cultivation; and through the water itself extends over only about 300 acres, it may be said to enrich five times that extent. (Dennison 1840, 366)

Dennison reported how 'after strong rains the washings of the streets and sewers of the town of Mansfield, which discharge to the River Maun, give great additional efficacy to the water' (Dennison 1840, 366). The system seems to have been extended in the middle decades of the century, for James Caird reported in 1852 that there were 400 acres (162 ha) of water meadows here. Cattle, kept almost exclusively in yards, were fed 'chiefly on the produce of the [water] meadows', so that 'sufficient manure is made to admit of an application of 30 tons to each acre' (Caird 1852, 205). It is thus not only the scale of the Clipstone meadows that is interesting, but the way in which the technology was being used in the context of the high-input, high-output farming systems of the High Farming period, with the grass and hay used to produce manure from cattle in yards, rather than to feed flocks of folding sheep. The period saw other novel ways of using irrigated meadows. At Dunster Castle in Somerset, for example, water descending from the hillside was used to irrigate, not permanent pasture, but two-year grass leys, in a rotation featuring wheat, turnips, and barley and two years of ley grass (Burke 1834, 531). Similar systems of mixed arable/grassland irrigation were also sporadically created in Scotland, as at Tealing in Angus (Fraser 2001, 135).

From the later eighteenth century long-established water meadows, as much as completely new ones, were generally provided with brick-built sluices and culverts, and many featured complex aqueducts designed to take carriers over the river to reach otherwise inaccessible areas. The meadows laid out at Castle Acre in Norfolk around 1808 included no less than sixteen brick-built sluices and two aqueducts (Figure 17) while a few kilometres upriver in East Lexham the canal-like main carrier – described on maps as the 'New River' – still crosses the river Nar on a massive, brick-built aqueduct (Wade Martins and Williamson 1994, 27–30). In Wessex, brick sluices and culverts seem to have proliferated, and increasing use was made of manufactured cast iron elements in hatches and sluices, produced by local engineering companies (Figures 19 and 20) (Cowan 1982; 2005). In all these ways, nineteenth-century meadows reflect the more industrialised and more capital-intensive agriculture of the High Farming period: and in the West Country, to the old habit of running carriers through farmyards and byres was now added the idea of using the water to turn waterwheels to power farm machinery, especially threshing machines.

Many long-established irrigation systems, especially in the west of England, were progressively extended and modified in the course of the nineteenth century, as at Escot, in the valley of the river Tale, in the South Hams district of Devon. A survey of 1763 described places on the estate where watering was already practised, or where it could be practised to advantage. Three meadows, covering 20 acres (8 ha), were described as 'Properly Draind under the Advantage of a water to be Turned over at pleasure'; another, the Eight Acres (actually covering 7.5 acres (3 ha)), was described as a 'meadow and proper for Meadow or Pasture but should not advise on Tillage here as its Under the same advantage of a water to be watered or Kept off, at yr leisure'. These meadows, probably bedworks in the areas 'C' and 'D' in Figure 21 (the former since levelled), were watered from a leat which ran down the western side of the valley, which also supplied water to the mill in nearby Ottery St Mary (later a wool factory). Some time before 1826, probably in the early years of the nineteenth century, another leat was added down the eastern side of the valley which irrigated areas on this side of the river ('A' in Figure 21), and also fed a waterwheel which was used to power a threshing machine at Clapperentale Farm. Subsequently, between 1826 and 1850, a second leat was added on the western side, and the course of the old leat here was modified. These changes were associated with the creation of a new area of meadows on this side of the valley ('B'), and almost certainly at the same time the existing areas of meadows ('A') on the opposite side of the Tale were comprehensively remodelled so that both now feature numerous well-built brick sluices and culverts, several of which have recently been revealed by excavations carried out by Exeter Archaeology on behalf of the Tale Valley Trust, who are keen to restore parts of this system to working order (Figure 22). These two areas ('A' and 'B') display a complicated mixture of catchworks (on the sides of the valley) and bedworks (on the valley floor).

The decline of water meadows

Water meadows thus have a complex history which extended long after the seventeenth century, and well beyond Wessex and the west. The late eighteenth and nineteenth centuries saw the expansion of floating into new areas, the use of the technique in novel ways within the agricultural economy, and the creation of more complex and sophisticated irrigation systems. All this was fuelled by the agricultural prosperity of the Napoleonic War and High Farming years, and by the development of an increasingly capitalised agriculture, of large tenant farms on large landed estates. It was also, to some extent, an expression of a fashionable interest in 'improvement' on the part of large landowners, itself in part the consequence of ideological and political as much as economic and agrarian concerns. Active involvement in agricultural innovation and land reclamation demonstrated, in the face of increasing radical opposition, that established landowners were the proper custodians of the countryside. Philip Pusey, describing in 1845 the potential benefits of irrigation in upland areas, commented:

> If the plain means of improvement and employment are still neglected, it will be possible to tax owners of needless deserts with supiness; and difficult to deny that they hold in their hands more of their country's surface than they are able to manage for their own good or for that of the community. (Pusey 1845, 521)

But, as is so often the case, the evidence of the landscape itself is most instructive on this issue. Many of the meadows described in this chapter lay on the home farms of large estates, very close to mansions and landscape parks – Clipstone, Escot, West Lexham, most of the Scottish examples. Others actually lay within parks, as at East Lexham or West Tofts in Norfolk, or at Woburn in Bedfordshire, where the Duke of Bedford's meadows were fed from the 'Temple Reservoir' (Batchelor 1813, 484–93; Wade Martins and Williamson

FIGURE 19.
The name of the engineer is proudly displayed on this derelict sluice in water meadows on the river Frome in Dorset.

FIGURE 20.
Nineteenth-century sluice in water meadows on the river Piddle at Bere Regis in Dorset.

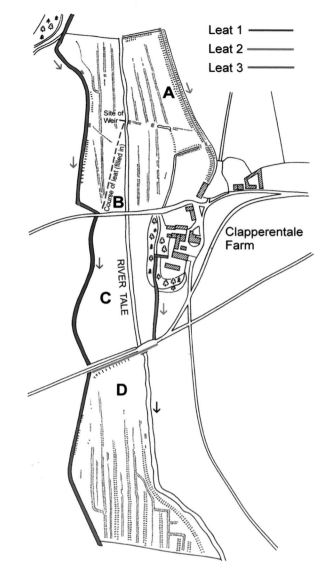

Leat 1 ———
Leat 2 ———
Leat 3 ———

Site of Weir

Course of leat (filled in)

A

B

C

RIVER TALE

D

Clapperentale Farm

FIGURE 21.
Plan of the water meadows at Escot in south Devon. The system, like many others, developed over a number of years (see text for explanation).

FIGURE 22.
Remains of a mid nineteenth-century brick-built sluice revealed by excavation in the water meadows at Escot.

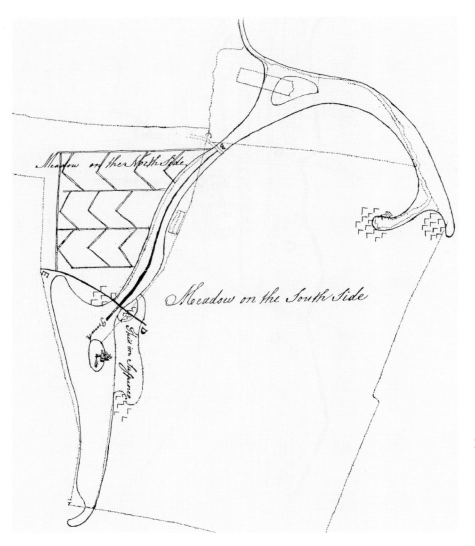

The words on the map read:

Meadow on the North Side

Meadow on the South Side

FIGURE 23.
Plan of water meadows
below the lake in the
park at Cusworth,
designed by Richard
Woods as part of his
'improvements' in
1762.

1994, 33–4). Indeed, the great Capability Brown himself created a number of irrigated meadows within the parks he designed, as did contemporary landscape gardeners like Richard Woods (DA BW/H/164; Figure 23).

Water meadows were still being managed with enthusiasm in many areas of Britain at the end of the nineteenth century. New books and articles on the technique continued to appear, including John Scott's *Irrigation and Water-Supply: A Practical Treatise on Water Meadows, Sewage Irrigation and Warping* of 1883 and Robie's 'On English water meadows and how far they are applicable to Scotland' (1876). Barker's *On Water Meadows as Suitable for Wales and other Mountain Districts* was still being republished as late as 1893, while MacDonald, writing in *Stephens' Book of the Farm* in 1908, described how 'great advantage' was still derived from the floated meadows in Wessex. But as the good years of High Farming gave way, in the decades after 1880, to a period of agricultural recession – which continued with only minor and short-lived interruptions up until the time of the Second World War – the days of water meadows were numbered. Watering first seems to have retrenched to its core areas, of Wessex and the West Country: by the early twentieth century most of the eastern examples had been abandoned, and few survived in upland districts. In the traditional heartlands, the decline was more gradual, but nevertheless steady, and by the 1940s the system was said to have 'broken down' even in Wessex (Stamp 1950, 81). Meadows were expensive to maintain at a time of falling prices, especially as the great folding flocks which they had helped to sustain became less important with the availability of cheap artificial fertilisers: in Wiltshire, for example, the numbers of sheep fell from 775,000 in 1870 to 162,000 in 1939. Moreover, the demand for early lamb was in decline from the 1880s as the advent of refrigeration allowed large-scale foreign imports, especially from New Zealand. Many farmers with extensive areas of valley-floor meadows switched their attention to dairy farming, and for a time irrigation played a new role in the farming economy, but reaper-binders, tractors and other heavy equipment damaged sluices, channels and bedworks, and got stuck in the mud. A variety of other factors contributed to the decline. Stamp, writing in 1950, noted how 'fishing rights have become more valuable and the owners of these rights may not approve the interference caused by the weirs; even worse a sportsman out for shooting may prefer to see a reed swamp than a tract of well tended meadow' (Stamp 1950, 81). In the upland areas of England, similarly, only a handful of catchmeadows continued to operate into the second half of the twentieth century, despite the continued advocacy of some agronomists. The widespread availability of other methods of improving hill pastures, including reseeding and nitrogen applications, rendered the old methods redundant.

CHAPTER SIX

Drowning by Numbers: The Functioning of Bedwork Water Meadows

Roger L. Cutting and Ian Cummings

Introduction

Water meadows have been described as areas of land in which 'deliberate controlled flooding was used to boost fertility' (RSPB *et al.* 1997, 6). Similar definitions recur in the literature, stressing that the flooding of land adjacent to watercourses promotes the growth of grass, which in turn provides early grazing for livestock and/or an increased hay yield. Yet at the same time the literature relating to the effects of flooding on soils and plant physiology suggests that such events are actually detrimental to productivity. Soils, when flooded, will become saturated (at a rate dependent on structure) and thereby oxygen-deficient and eventually anaerobic. Such conditions should be detrimental to grass growth, both in terms of the soil chemistry through the loss of available nitrate and the production of phytotoxic substances (Cook 1999), and because of a drastic diminution of oxygen supply to the roots (anoxia) (Kramer and Boyer 1995, 193–99). Tolerances to anoxia vary but many plants will die within a few days of flooding (Brix and Sorrell 1996). 'Floated meadows' thus appear to contradict the normal rules of soil chemistry and plant growth: they are deliberately irrigated for weeks at a time throughout the winter months, yet not only do the grasses survive such severe environmental and physiological stress, but apparently thrive.

It is important, however, to stress the distinction between the simple flooding of meadows and the technique of 'floating', for the latter involves the rapid flow of water across the land surface. As explained elsewhere in this volume, this was achieved either by utilising the local topography (in the case of catchwork meadows) or by creating artificial ridges, and allowing water to be deployed via channels cut along their apex: the channels were allowed to overflow and the water cascade down the slopes to collect in drains and so eventually be returned to the river.

How such systems promote grass growth is explained in some detail in the abundant historical literature concerning floated water meadows. Two main factors are regularly cited. Firstly, the water warms the soil, thereby protecting it from frosts and promoting early growth (Wright 1790). This improvement in productivity would in turn require a nutrient supply and this is allegedly provided by the second factor, namely the supply of so-called 'dirty water' (a contemporary term relating to river water containing high levels of effluent) that would in some way fertilise the soil and so allow vigorous growth to be sustained.

Traditional explanations do not, however, fully clarify the processes by which productivity is promoted. For example, early growth is required in February, yet it was usual practice to drown the meadows throughout the winter, putting the grass under a considerably greater physiological stress than the winter frosts themselves. Furthermore, if warming the soil was a primary objective then the timing and frequency of floating is curious: recent work has shown that the soil temperatures of floated meadows equilibrate to those of the irrigation water within approximately three to four days (Cutting *et al*. 2003). In terms of raising the temperature of the soil, in other words, it would have been just as effective to float the meadow over the course of a few days in early February, as it would have been to do so for sustained fortnightly periods throughout the winter months.

The second feature of traditional floating, namely the use of 'dirty' water, also raises a number of questions. The literature often stresses the need for the water to be as 'turbid, feculent and replete with putrescent matter' as possible (Wimpey 1786, 295), this being seen as most effective as a fertiliser and conditioner of the soil. But the nature of the suspended materials is important here. Fluvial sediments will often include quantities of organic and partially decomposed organic materials, the introduction of which increases the amount of carbon in the meadow soils, significantly altering their carbon–nitrogen (C/N) ratio. The decay of such materials requires an increase in the bacterial, fungal and actinomycete populations, and this in turn significantly reduces available nitrate nitrogen, which is instead required to build tissue for the growing microbial populations. During this so called 'nitrate depression period', nitrogen available in any form for plant uptake that may be present in the soil all but disappears (Brady 1974).

Early writers on water meadows often emphasised the need for a continuous flow of water. This is also curious, because once the soil reaches saturation, regardless of the nature of surface flow, all available pore spaces are filled with water and the resulting deficiency in aeration reduces root development and, in consequence, the absorption of nutrient minerals is reduced. In other contexts, such nutrient deficits associated with flooded soils are generally harmful to plants (Kramer and Boyer 1995, 193–9).

Floated water meadows are thus intriguing. They were once a common feature of the countryside, especially in the river valleys of southern England, and they were undoubtedly highly productive: yet they ought not to have been. Indeed, our present understanding of soil systems may even suggest that the reverse should have been the case.

Addressing these issues (as well as elucidating a number of others in relation to meadow productivity) was the primary aim of research conducted at the Britford Water Meadows (SSSI) and the Lower Woodford Water Meadows (SSSI), Wiltshire, from 1997. Both of these sites represent rare examples of meadows that are still actively floated. The Lower Woodford system covers approximately 24 ha and the Britford system 18.2 ha. Both are situated on the Wiltshire Avon, respectively 8 km to the north, and immediately to the south, of the city of Salisbury (SU 166274 and SU 123347). Both systems have a somewhat haphazard appearance, suggesting piecemeal construction over a period of time, but they undoubtedly represent nearly 300 years of uninterrupted management (Bettey 2003).

Temperature and the 'hot bed'

Protecting the soils from winter frosts is frequently cited as one of the most important objectives of floating as an agricultural management technique. Much has been made of the high temperature of emergent spring water, and both the historical (Smith 1856a) and the modern literature suggest that this would serve to warm the soil, maintaining the temperature above freezing and therefore allowing the grass to grow much earlier in the season than would otherwise have been the case.

FIGURE 24.
Results of hand-held temperature readings made between 1997 and 1998 at the Britford water meadows. Points represent mean values for temperature (°C), n=100. Error bars represent +/- 1 SD from the mean.

Soil temperature regimes are complicated by the fact that they will normally exhibit two fluctuations, one seasonal in character and the other diurnal. Both of these, moreover, vary with depth. The greatest variation may be expected on the surface, with the range being dampened exponentially with depth, the diurnal fluctuations at 350 mm being only about 5 per cent of those at the surface (Russell 1973). Any influence of temperature on growth through floating must, therefore, take place near the surface.

Soil temperatures were recorded throughout the winters of 1997/8 and 1998/9 at the Britford water meadows at a depth of 200 mm, using an ELE MM900 multi-channel environmental monitoring station (EMS). Air temperatures were also recorded, together with water temperatures during periods of flooding. The EMS was configured to scan every hour and to log four-hourly means. Eight channels were dedicated to soil temperatures; four were placed in efficiently watered areas and four in an adjacent unfloated 'control' area. The monitoring station was relocated to the Lower Woodford site in the winter of 2002/3. The static location of the probes, combined with the discretionary management of these water meadows, meant that the data needed to be verified by handheld soil temperature readings: in all, 2,300 spot determinations of soil temperature in both floated and unfloated areas of the Britford water meadows were collected during the winter period.

Analysis of these data revealed that once an area was flooded, it would take approximately three days for the soil temperature to equilibrate to that of the surface water (Cutting *et al.* 2003). At this point soil temperature fluctuations would closely parallel that of the flood water (although at Lower

Woodford significant heat loss was recorded as the water flowed through the sward on the side of the beds and into the adjacent drains (Cook and Cutting in preparation)).

Figure 24 shows the relationship between air temperature and soil temperatures on floated and unfloated sites within the Britford system. A clear mean difference is apparent but, at only 2.5–3 °C, it is somewhat less marked than much of the historical literature might lead us to expect. Verification of these results came from the Lower Woodford site: here, while it was evident that floating had a clear effect on soil temperature, the difference between floated and unfloated sites was only 2.5 °C. Further analysis of the data suggested that at higher average air temperatures (above 10 °C) the soil temperatures of floated and unfloated sites converged. Indeed, at the highest air temperatures there was no significant difference between the floated and unfloated soils. As the air temperatures fell, so the significant difference was

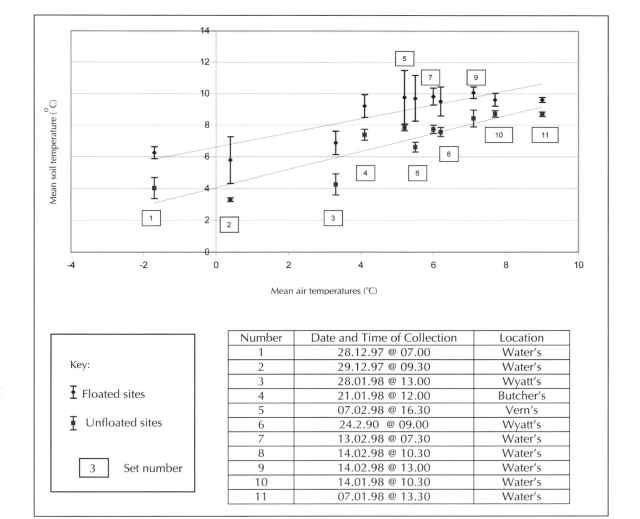

Number	Date and Time of Collection	Location
1	28.12.97 @ 07.00	Water's
2	29.12.97 @ 09.30	Water's
3	28.01.98 @ 13.00	Wyatt's
4	21.01.98 @ 12.00	Butcher's
5	07.02.98 @ 16.30	Vern's
6	24.2.90 @ 09.00	Wyatt's
7	13.02.98 @ 07.30	Water's
8	14.02.98 @ 10.30	Water's
9	14.02.98 @ 13.00	Water's
10	14.01.98 @ 10.30	Water's
11	07.01.98 @ 13.30	Water's

Key:

$\underline{\mathbf{I}}$ Floated sites

$\underline{\mathbf{I}}$ Unfloated sites

3   Set number

re-established. This implies that the flood water does not necessarily warm the soil, but rather restricts the amount of heat loss during periods of cold weather, and at night.

A complicating factor in the relationship between temperature and plant growth is the time-lag involved. Temperatures may change rapidly over a short period of time but the physiological response of the grasses will, of necessity, be delayed. It has been suggested (Alcock *et al.* 1968) that crop yield is highly correlated with 'accumulated day degrees' of soil temperature. In other words, soil temperatures need to be maintained over a period of time before they have a significant and measurable effect upon growth. A threshold temperature of 5.5 °C was suggested for rye grass (*Lolium perenne*), a dominant species in the Britford water meadows. Peacock (1975) has determined that growth of *L. perenne* will take place at temperatures below 5 °C, but that a clear and consistent exponential relationship exists between leaf extension and the temperature through a range of between 1 °C and 10 °C. The time period for which the soil would need to be maintained at or above this temperature would be ten days, after which point significant growth occurs.

The mean temperatures of the floated soils at Britford and Lower Woodford were largely maintained above 5.5 °C throughout the winter. Such management, sustained for a two-week period, would therefore easily reach the threshold for rapid grass growth. Despite the floated soil temperature reaching that of the irrigation water within four days, the requirement for a sustained temperature at or above 5.5 °C perhaps explains the traditional preference for a longer floating period. Sustained ten-day periods of drowning over the winter months maintains soil temperatures well above the required threshold, thereby establishing one of the necessary environmental parameters for early growth.

In addition, when the hatches were opened and the water removed from the meadows, a heating lag effect was consistently identified at sites across the Britford system. The soil temperatures of the floated areas took between three and four days to equalise with those recorded for the unfloated control sites. This affords the floated sites a period of time during which soil temperatures are maintained above the threshold of 5.5 °C accumulative day mean temperature for the duration of the drowning period and, despite the rapid re-establishment of aerobic conditions, the soils remain relatively warm, promoting conditions suitable for early growth.

The requirement for dirty water

Despite some debate (Marshall 1796a, 209 and Wright 1799, 24), most early writers on water meadows believed that so-called 'dirty water' produced the greatest quantity, and the best quality, of grass. Indeed, in Boswell's famous treatise on floating emphasis is actually placed on water quality to the exclusion of any reference to considerations of soil temperature (Boswell 1779). Given the imprecise nature of the term 'dirty water' it is assumed here that it simply referred to river water in which dissolved and organic and inorganic suspended sediments were present. To elucidate the relative importance of waterborne sediments a numerical model of sediment and nutrient fluxes was created from soil and water samples taken at strategic points across the Britford water meadows. The sample frame essentially followed the passage of water from the main carrier, to smaller carriers, to the drains and eventually to the tail drain. The water samples were analysed for the primary macro-nutrients required for plant growth: nitrogen and phosphorous. Soil samples were also analysed for their nutrient status and texture.

The timing and the method of the initial flooding is a critical factor as during drowning, once a steady state has been established, the extracted water runs clear. Indeed, in relation to the suspended sediment, a mean concentration of 0.042 g l^{-1} was recorded coming on to the meadow and 0.04 g l^{-1} returning to the tail drain. These low levels may be expected when the water is drawn from the chalk-fed streams of south-west England. However, a significant input of sediment takes place within the first hour after the opening of the hatches. A seemingly vital process described by Wright (1790, 31) and again by Browne (1817, 26), but generally overlooked in the modern literature, is the initial disturbance of the deposited sediments in the stream or river. Indeed, Wright goes so far as to advocate the raking of the river bed immediately prior to floating. At Britford and Lower Woodford the hatches are opened and closed in quick succession, thereby creating disturbance through rapid displacement leading to significant entrainment, and consequently a marked increase in turbidity. Only when the water has become turbid and charged with sediment is it allowed to flood the meadows. The initial high-energy conditions prevalent as the water fills the carriers may also be expected to entrain further fine sediments. Mean sediment concentrations recorded at Britford during initial flooding were in the range of 1.5 g l^{-1} to 2.5 g l^{-1}. Readings taken at the drains as the initial flood water arrived, having crossed the system, recorded suspended sediment masses in the range of between 0.02 g l^{-1} and 0.4 g l^{-1}, implying that approximately 98 per cent of the sediment had been deposited within the water meadows. Within one hour, however, the suspended sediment had returned to a range similar to that of steady-state conditions, thus reinforcing the importance of this initial input.

During steady-state conditions Nitrate-Nitrogen (NO_3-N) is the dominant N species in the irrigation water, reflecting its high oxygen status. The surface water showed no significant reduction in NO_3-N, or in total oxidised nitrogen concentration, across the system during floating. Supporting this finding, a ^{15}N tracer study suggested that uptake by the sward was as little as 1 per cent of the NO_3-N applied during irrigation. Indeed, analysis of the soil water during floating showed a significant decrease in mean topsoil NO_3-N. This increased as water was removed only to decrease again with uptake. These findings suggest that there was leaching, mineralisation and nitrification followed by uptake, in parallel with the drowning cycle.

Phosphorous (P) is rarely found in high concentrations in river water because it is rapidly utilised by plants. It normally occurs as dissolved ortho-phosphates, polyphosphates and organic phosphate (i.e., that which is already organically bound), and continuous changes occur between these phases as P is taken up, synthesised and then liberated through decomposition. Tracking the movement of P through a system is complicated not only by these transitions, but also by the variable rates of uptake across the year, the availability of P being much reduced during growing periods. As only small amounts of P are directly dissolved in water as ortho-phosphate, much larger amounts of P may be introduced into a wetland system by being incorporated into the sediments transported as the suspended load. The amount of sediment entering the water meadows during floating will therefore be related to the input of total P.

During the initial drowning period at Britford samples were collected from two points adjacent to main hatches and from these it was estimated that between 383 kg and 498 kg of suspended sediment was introduced via the carriers during the first hour of irrigation. Assuming that all the sediment sampled from the bottom of the navigation was mobile, then it becomes possible to calculate the weight of available P and total P entering the meadows. After this first hour, it is assumed that sediment and P fluxes are conservative as the meadow approaches steady-state (Cutting *et al.* 2003).

The first hour provided excess suspended load of 363 and 617 g of total P over steady-state entering the meadows from the two sample points respectively. Assuming a typical vertical infiltration rate at steady-state of 7 mm h^{-1} and a concentration of 0.16 mg l^{-1}, the soil solution of the topsoil received 11.2 g ha^{-1} h^{-1} of dissolved P in excess of that trapped from the suspended load. However, initial infiltration rates would be greater at the outset of floating, making the instantaneous intake of P in the topsoil large.

The distribution of the suspended load entering the meadows is unlikely to be even because of the trapping of silt and sand grade material during the passage of the water through the grass (Cook *et al.* 2004). Sediment and topsoil samples were taken in the summer of 2000 when the meadow had not been floated for four months. These were collected from strategic locations along the hydrological pathway representing entry, mid-meadow and exit points. A significant difference between the 'upper system' samples and the 'lower system' was found, suggesting that passage across the meadows more than halves the mean available P (from 122 mg kg^{-1} to 59.5 mg kg^{-1}). Differences in mean total P were insignificant. When dry, the system showed high levels of available P compared with winter-sampled soils and sediments.

Samples were taken between irrigation events and these again showed a fall across the system in available P, by about one-third. Furthermore, topsoil samples collected during drowning also showed significant differences in available P between the upper and lower systems. These results leave no doubt that the passage of water across the meadows leads to a significant deposition of sediment, and that there is an uneven distribution of available P.

Surface flow

The soils of the Avon Valley floodplain are mapped by the Soil Survey (Findlay *et al.* 1984) as falling within the Frome Association – typically silty clay loams with a substrate of calcareous gravel at about 300 mm depth. Being ancient pasture land the topsoil normally exhibits a good crumb structure, imparting a high porosity. For these soils to maintain a saturated overland flow requires a significant input of water. Initially this input is required to fill pore spaces within the soil, thereby allowing surface flow to be maintained. After this the flow needs to be maintained to a depth of up to 30 mm for several weeks at a time.

Initially, soil infiltration rates at Britford were calculated in order to ascertain the rate at which the profile would reach saturation. Typical initial rates on these silty clay loams were between 30 to 40 mm min^{-1}, slowing to 10 to 20 mm min^{-1} after ten minutes. These high initial rates suggest rapid saturation of the topsoil, but soil water would need to pass continuously through the soil in order to avoid the establishment of anaerobic conditions. The Auger Hole Method (van Beers 1958) was employed to gauge the saturated hydraulic conductivity (the lateral permeability) within the underlying gravels, which in this area are held in a tightly packed sandy loam matrix. Initial rates of 20 mm day^{-1} were recorded, which may be classified as slow (Thomasson 1975): yet such permeability does allow the lateral movement, and thereby replacement, of soil water within the profile.

Having established infiltration rates, the saturation of the soil profiles was ascertained using a field capacitance probe. Monitoring in this way during drowning suggested that the topsoil approached saturation within five minutes of the onset of drowning. Field observations suggested that sections of the Britford water meadows were efficiently operating within seventeen minutes. Likewise, when the hatches controlling the movement of water on to the meadow were closed and the water removed, cessation of surface flow was recorded after seven minutes, and the soil was firm underfoot within an hour.

In terms of surface flow the Britford water meadows represent an approximation to a closed hydrological system, in that the water comes on to the meadow from canalised channels and returns to the river via the tail drain. Despite some limited external inputs from precipitation an estimation of the total amount of surface water entering and leaving the system can therefore be made. The ingress of water entering the meadow system was manually gauged, as was the discharge into the tail drain. Because the Britford water meadows comprise four discrete compartments, a number of readings could be taken. Furthermore, within each compartment, smaller areas are supplied by specific carriers, and discharge into particular drains. By taking a range of measurements across the various meadows, each of different area, it proved possible to estimate (on the basis of data collected over a period of two years) not only the total mass budget of water but also the discharge of water required to float a given area of grass.

The plant communities identified at the Britford water meadows (Cutting and Cummings 1999) experience long periods of sustained flooding. Higher plants are obligate aerobes which require oxygen for a range of physiological processes, and such flooding events will inevitably reduce gas exchange, particularly within the rhizosphere. Although many species may experience anoxia at some time during their life cycle, plant tolerance to such events varies from a few hours to several weeks (Crawford 1989, 105–29). The way plants tolerate such conditions and the adaptive strategies they adopt are discussed elsewhere (see Chapter Seven). However, the maintenance of flood conditions would normally lead to reducing chemical conditions induced by the depletion of oxygen, with the resulting establishment of an increasingly hostile biochemical environment. To achieve the enhanced productivity that these meadows display (Cutting and Cummings 1999) there must, therefore, be a mechanism by which the establishment of severe chemical reduction is avoided during flooding. The continuous movement of water across the surface, frequently stressed in the historical literature, appears to be crucial here.

As the carriers overfill, the water runs down the panes in a continuous sheet of water which ideally has a depth of between 25 and 30 mm. The precise character of the movement appears to be critical here: the water flows *through* the grass, rather than overtopping it and flattening the sward (Figure 25). In this

FIGURE 25. When bedwork water meadows are 'floated', the water flows *through* the grass, rather than overtopping it and flattening the sward, so that it is forced into highly turbulent patterns of flow.

way the water is forced into highly turbulent patterns of flow. This increases the surface area of the water across which atmospheric oxygen can diffuse. Determinations of dissolved oxygen in surface water at Britford returned mean values in a range of 85 to 92 per cent saturation. This compares with readings taken in standing flood waters, in control sites adjacent to the meadows, of from 54 to 71 per cent saturation.

Infiltration of the highly oxygenated surface water is reflected in the dissolved oxygen levels of soil water. Readings taken from replicated clusters of piezometers (soil water samplers) suggested high levels of dissolved oxygen in the water entering the soil. This highly oxidative environment evidently prevents the establishment of biochemical reduction and the associated detrimental effects on plant growth. It may be assumed that without the constant flow of water, anaerobicity would be established.

Productivity

To establish the impact of floating on grass productivity, plots 1 m² were established as part of a replicated block design at the Britford water meadows, in both floated and adjacent unfloated sites. These plots were cropped weekly immediately after the water had been removed for the last time. Increases of between 60 and 80 per cent in early-season dry-matter production were recorded during the five weeks after floating. Early-season floating of water meadows certainly brings forward the productivity of grass in quantities that would have been economically worthwhile before the twentieth century. It remains uncertain, however, whether year-round productivity is affected. Attempts to verify these field data were carried out at the Rural Technology Unit (RTU) at the University of East Anglia, Norwich, UK. Experimental plots were established and water extracted from a local borehole was run across them to simulate floating. The difference between the floated and unfloated plots is minimal during irrigation, but in the weeks following the cessation of floating it became increasingly marked. The increases in early productivity in the experimental plots were comparable to those measured at the Britford water meadows. The RTU experimental plots were floated for three consecutive seasons. Interestingly, productivity declined year on year between 1998/9 and 2000/1. In the final year that statistically significant differences were established between floated and unfloated plots, productivity had halved, suggesting that for ungrazed systems, with no application of fertilisers, such a management protocol is unsustainable in the medium to long term, an observation which clearly serves to emphasise the importance of nutrient inputs from the river water.

The operation of floated meadows

The field research presented allows the formulation of a model which can describe how, and why, water meadows work (Cook *et al.* 2004). In summation, the process begins with the disturbance of the water prior to floating, which entrains sediment just before the carrier hatches are pulled. Once the hatches are opened, the heavily sediment-laden water, its competence increased by the nature of flow and high velocities achieved, rapidly fills the carriers. These are quickly overtopped and the adjacent soils wet up, swiftly reaching saturation and leading to the onset of overland flow through the sward. The flow through the grass is necessarily turbulent, and thus oxygenates the water. Slow infiltration and saturated hydraulic conductivity within the substrate allow the continued, albeit slow, permeation of highly oxygenated water, thereby avoiding chemically reducing conditions. Grasses may also survive through a number of adaptive strategies, as well as through long-term selection (see Chapter Seven).

The soil temperatures converge to the water temperature within days, and the water prevents heat loss from the soils during cold periods of weather, thereby maintaining the soils at the accumulative day temperature of 5.5 °C. During irrigation some limited grass growth does take place and would necessarily require nutrients, but there is little evidence that water meadows act to reduce the concentration of inorganic nitrogen in the surface water. Any nitrogen removed may well be replaced by that leached from the soil. It is suggested that meadows do, however, act as effective traps for suspended sediments, and therefore for the associated P.

With the cessation of irrigation, the soils rapidly drain and aerobic conditions are quickly re-established. Furthermore, the warmth provided by the water is maintained for three to four days, and the soils can also be assumed to be near field capacity in relation to water availability. Aeration, relatively higher temperatures and available water are conducive to microbial activity, which in turn plays an important role in oxidation, mineralisation and nitrification. Grass growth is triggered by the rise in temperature and the sward is able to utilise inputs of nitrogen, and of newly introduced phosphorus from the irrigation water. Each of these physical and biochemical processes appears important in the promotion of early productivity; their relative importance, however, depends perhaps on the physical nature of the site. Nonetheless, it does appear that it is the effective integration of these mechanisms which was crucial to the long-term sustainability of water meadow systems.

A future for water meadows

Present-day research can thus provide historians and others with much-needed empirically validated insights into this traditional form of management, explaining much about the way floated meadows operated, and about how precisely they served to increase grassland productivity. The technique of floating clearly provides an effective stimulus for grass growth at levels which compare favourably with modern, intensively managed and commercially treated pastures. Water meadows represent a sustainable grassland production system which is environmentally benign and which is still, in a few locations, practised on a commercial basis. Of particular interest is the way that the Britford water meadows receive irrigation water high in N and P, typical of many valleys in southern England – particularly those in which agricultural diffuse loading and sewage effluent discharges occur. It appears that meadows play a role in trapping sediment, and in reducing the amounts of soil-available P. Such transfers from an aquatic habitat to a terrestrial grazing system suggests that other diffuse pollutants may follow similar pathways, although this is a matter which requires further research. In the bio-assimilation of nutrients, the diffusion of heat and the promotion (when operated effectively) of productivity, floated water meadows not only form an important part of our historic landscape heritage but also provide us with a technique from the past which may very well have a number of future functional applications to a range of urgent contemporary issues.

CHAPTER SEVEN

The Effects of Floating on Plant Communities

Ian Cummings and Roger L. Cutting

Introduction

Water meadows are semi-natural floodplain habitats which are important for a wide range of native species of plants and animals. Uninterrupted management for over 400 years has resulted in recognisable plant communities, assemblages of species which are rarely found in the wider countryside. Originally designed to increase agricultural production (Cowan 2005), water meadows are today increasingly being appreciated for their environmental and ecological value. The future of these intriguing systems is, however, in the balance as pressures on floodplains continue to increase. In order to inform the debate about their potential use and/or conservation, it is necessary to provide a detailed ecological evaluation of these sites. This chapter focuses on the plant communities of two representative examples of water meadows in the United Kingdom, one still managed by floating or drowning, the other now abandoned.

The sites in question are the Britford water meadows SSSI near Salisbury, Wiltshire (SU 166274); and the Castle Acre water meadows, Norfolk (TF 824151), which lie within the River Nar corridor SSSI. Whilst differing in many respects and located in different parts of the country, the two meadows provide an interesting comparison in that both are of bedwork construction and support vegetation communities which, while superficially different, display a number of underlying similarities.

The Britford system is still actively floated (Cutting and Cummings 1999; Cutting *et al.* 2003) and is described in more detail elsewhere in this book (pp. 102–4 and 122–8). The site lies on soils of the Frome Association, being typically grey, mottled, silty clay loams overlying calcareous gravel (Findlay *et al.* 1984). The Castle Acre water meadows, in contrast, were abandoned in the late nineteenth century after a relatively intense period of operation from 1806 to around 1890 (Wade-Martins and Williamson 1994). The broad outline of the main channels can still be clearly distinguished on the 1946 RAF vertical aerial photographs (Figure 26). The soils here are classified by the Soil Survey as falling within the Isleham 1 Association (Hodge *et al.* 1984): typical

FIGURE 26.
Vertical aerial
photograph of the
Castle Acre water
meadows, Norfolk,
taken by the RAF in
1946.

humic-sandy gleys predominate throughout the site, with a lithology of sandy and peaty, glacio-fluvial drift and fen peat. The flora at both sites was exhaustively surveyed, and some of the hypotheses resulting from these observations were tested by an experiment at the Rural Technology Unit (RTU) at the University of East Anglia (TG 198070) between 1998 and 2001.

Systematic botanical records of the flora of water meadows exist from the Britford area from as early as 1610, and in 1863 Dyer described the exceptional hay yields brought about by floating:

> At Orcheston St Mary, about eleven miles from Salisbury, is a small tract of meadow land, half a mile from the village of Shrewton which is sometimes watered in the winter by means of a small spring flowing out of a limestone rock. It is mown thrice in summer and after a favourable season for watering, the first crop is nearly five tons per acre, the second about half as much. (Dyer 1863; 1864)

Dyer drew particular attention to the extraordinary length of some of the meadow grasses, recording a 'yearly growth of 10 foot long and commonly sixteen foot long in irrigated meadows' (Dyer 1864).

This astonishing productivity excited the attention of the Agricultural Society established at Bath in the late nineteenth century, and their investigations revealed that the principal grass in the hay crop was Creeping Bent, *Agrostis stolonifera* (nomenclature follows Stace 1992). It is possible that this and similar observations influenced the kinds of seed mixtures which nineteenth-century writers recommended for establishing new water meadows. The *Encyclopaedia Britannica* thus recommended in 1880, on the basis of 'considerable experience', that meadows should be sown at a density of 46 lb per acre (52.2 kg per hectare) with a mixture of *Lolium perenne*, *Poa trivialis*, *Glyceria fluitans*, *Glyceria aquatica*, *Agrostis stolonifera*, *Festuca elatior*, *Festuca loliacea*, *Phleum pratensis*, *Phalaris arundinacea* and *Lotus corniculatus*. To what extent this particular mixture was widely adopted is not known, but this in a sense perhaps matters little, for evidence is beginning to emerge that the practice of floating tends to select certain plants over others, producing recognisable communities regardless of what was initially sown in the ground.

In recent years, detailed ecological studies have indirectly tested historical observations and revealed that certain plants respond in unusual ways to floating. In a number of experiments it has been shown that several species common in water meadows respond in similar ways, notably with increased stolon length, increased dry weight production and an ability to grow under oxygen deficient conditions (Rozema and Blom 1977; Davies and Singh 1983; Visser *et al.* 2000). This chapter will review some of the effects of floating on meadow species and the types of plant communities that arise from the long-term practice of the technique.

One of the most noticeable features of the Britford and the Castle Acre meadows is the way that, in spite of differences in soils and current management, both are dominated by one main plant community: that classified by the National Vegetation Classification scheme (NVC) as type MG11 (*Festuca rubra – Agrostis stolonifera – Potentilla anserina*: Rodwell 1993; Cutting and Cummings 1999). This is widely recognised as a relatively species-poor grassland community, and it is thus superficially surprising that both sites carry the designation of Sites of Special Scientific Interest, SSSI. The explanation lies in the rich and diverse communities found in the rivers, on their banks and in the ditches draining the meadows: communities which include S5a, *Glyceria* swamp; MG8, *Cynosurus cristatus – Caltha palustris*; MG9a, *Holcus lanatus – Deschampsia cespitosa*; and S14c, *Sparganium erectum* swamp (Rodwell 1991; 1995) (plant community analysis was undertaken using *Tablefit* (Hill 1993, table 1)).

The main difference between the two sites lies in the relative proportions of each community. Whereas the majority of the Castle Acre water meadow is classified as NVC-type S5a (*Glyceria maxima* swamp), at Britford this community is confined to the drains and ditches, the greater part of the meadow consisting of high-yielding species-poor MG11 grassland (Table 1). At first glance, the lush growth of marsh plants between the abandoned bedworks at the Castle Acre site is striking; dominant plants include *Filipendula ulmaria*, *Typha latifolia*, *Glyceria maxima* and *Cardamine pratensis*. Closer observation, however, reveals the underlying similarity between the two sites and it is tempting, and not unreasonable, to speculate that the Castle Acre water meadows were once more akin to those at Britford than they are today, having reverted to a riparian marsh community when floating ceased in the late nineteenth century.

Sampling location on pane	Water's Meadow NVC type	Butcher's Meadow	Drains in Butcher's Meadow
1	MG9a	MG11	
1	MG11	MG11	
1	MG11	MG11	
1	MG11	MG11	
1	MG11	MG11	
1	MG11	MG11	
2	MG11	MG11	
2	MG8	MG11	
2	MG11	MG11	
3	MG11	n.d.	
3	MG11	S5a	
3	MG11	MG11	
4			S5a
4			S5a
4			S14c
4			S5a

Table 1. A comparison of the NVC (National Vegetation Classification) plant communities recorded at Britford water meadows, Wiltshire (SU 166274) on (1) upper, (2) middle and (3) lower positions of the panes in Water's and Butcher's Meadows and in the drains (4) surrounding Butcher's Meadow; n = 3 – 6; n.d. = not defined. Data analysed using *Tablefit* (Hill 1993); NVC communities according to Rodwell (1991, 1993, 1995); adapted from Cutting and Cummings (1999): MG11 = *Festuca rubra – Agrostis stolonifera – Potentilla anserina*; S5a = *Glyceria maxima* swamp; MG8 = *Cynosurus cristatus – Caltha palustris*; MG9a = *Holcus lanatus – Deschampsia cespitosa*; S14c = *Sparganium erectum* swamp.

Experimental evidence for high grass yields

In an attempt to throw light on sward yields, as well as the character of vegetation communities, of water meadows Cook *et al.* (2004) measured the weekly dry weight of vegetation cropped from three separate meadows at Britford (Figure 27). Two meadows (Verne's and Butcher's) had been recently floated; a third meadow was unfloated, and served as a control. In four successive weeks, both of the floated meadows produced significantly higher dry weights than the control. Five days after floating Butcher's thus displayed a mean productivity of 22.5 g/m², roughly twice that of the control at 11.6 g/m². After a month

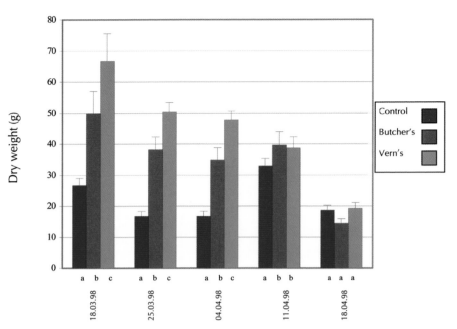

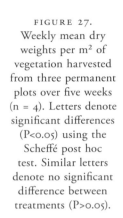

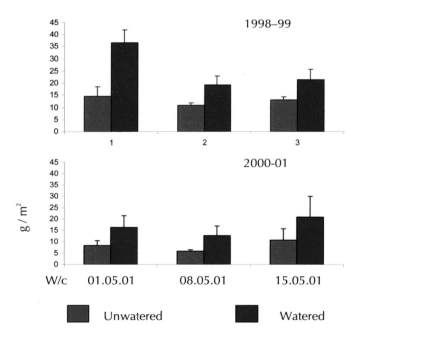

FIGURE 27.
Weekly mean dry weights per m² of vegetation harvested from three permanent plots over five weeks (n = 4). Letters denote significant differences (P<0.05) using the Scheffé post hoc test. Similar letters denote no significant difference between treatments (P>0.05).

FIGURE 28.
Mean dry weights (g) harvested from watered and unwatered treatments at the Rural Technology Unit at the University of East Anglia, Norwich, in 1998–9 and 2000–1.

the differences between the two areas was not significant. Nevertheless, the difference between early and late season productivities compares favourably with nineteenth-century estimates of 5 tons/acre (12.7 tonnes/ha) after the first mowing (roughly two months after floating), dropping to 2.5 tons/acre (6.4 tonnes/ha) for the second.

FIGURE 29.
Comparison of the effects of floating on grass and non-grass species on mean dry weight ± 1 SD per 1 m² (n = 4). Plots were cropped for a four-week period at the Rural Technology Unit, University of East Anglia, from 17 May to 7 June 2001.

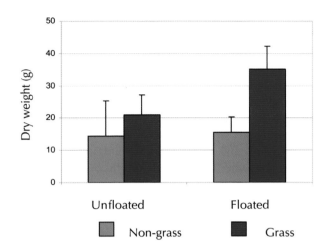

These observations were tested on an artificial water meadow established at the RTU at the University of East Anglia, arranged in a randomised block design incorporating 'watered' and 'unwatered' controls. Results were broadly consistent with the patterns observed at Britford, with significantly higher yields in watered treatments than in the controls (Figure 28). A further observation from this study was that when vegetation was separated into monocotyledenous ('grasses') and dicotyledenous species, floating produced a significantly greater dry weight in the former: in the unfloated controls differences between these two groups were not significant (Figure 29, after Cutting 2002). These results imply two important things: firstly, that floating does indeed increase the productivity of grassland; and secondly, that plant species respond in various ways to the treatment, leading to the dominance of the monocotyledonous species.

Inundation experiments on individual species

Many experiments have been conducted in the past on the effects of flooding on individual species, but few have examined its impact on species *mixtures*. Two deal specifically with the effects on species associated with the MG11 community found at Britford. Rozema and Blom (1977) performed a classic experiment on two species associated with periodic inundation: *A. stolonifera* and *Juncus gerardii*. Table 2 summarises a selection of the results. Plants were grown in trays for six weeks and subjected to three different treatments, reflecting a range of different levels of flooding: water table maintained 5 cm above the surface; at the surface itself; and at 'field capacity' (roughly equal proportions of soil and air by volume – equating to a water potential of about –5 kPa). At the end of the six-week experiment, *A. stolonifera* responded to the high water table by growing *stolons* – that is, aerial shoots with the ability to produce adventitious roots and new plants which may then become independent (called '*ramets*') of the original plant (referred to as the '*genet*').

mechanism was suggested above for *J. gerardii*, and evidence for homogeneous swards of well-adapted species can be seen at Britford. Alternatively, some species appear to have special adaptations which seem to fall into two main camps: firstly, plants that root superficially – taking advantage of the oxygen available in the upper few millimetres of soil (e.g. *A. stolonifera*, seen in abundance at both Britford and Castle Acre, and at other water meadows); and secondly, the development of *aerenchyma* – specialised 'air cells' that maintain a constant supply of oxygen to roots and submerged stems (Visser *et al.* 2000).

As reducing conditions increase in the soil with increased levels of inundation, a number of chemical changes take place, such as the formation of the cations ammonium (NH_4^+), Manganese (Mn^{2+}) and Iron II (Fe^{2+}), and of the Sulphide anion (S^{2-}). Whilst plants can acclimate for short periods, eventually aerobic respiration declines. Whilst all plant species synthesise anaerobic proteins that enable oxygen-independent ATP production, well-adapted species use fermentation pathways to survive under water for many months (Subbaiah and Sachs 2003). However, in crowded grassland communities impaired growth will limit the competitiveness of less well-adapted species and may result in their decline. Amongst one of the better-studied effects of flooding is the production by submerged tissues of the hormone ethylene, a potentially phytotoxic gas (Visser *et al.* 1997). Plants that possess (or can produce) aerenchyma have a distinct advantage over those that cannot, since they can vent toxic gases such as ethylene and methane and therefore survive the major physiological stress imposed by floating.

Aerenchyma and Radial Oxygen Loss

Aerenchyma facilitate the transport of gases from aerated shoots to organs surrounded by anaerobic conditions, and are of two main types. Those that form through cell collapse are known as *lysigenous* (Justin and Armstrong 1987). They are found in taxa such as *Nymphoides*, *Littorella*, *Luronium* and *Juncus*, including many monocotyledonous species (*Agrostis* and *Festuca*). The other type of aerenchyma are known as *schizogenous* and are found in *Caltha*, *Rumex*, *Filipendula* and many dicotyledenous species. Constitutive aerenchyma (present with or without flooding) are found in *Juncus effusus* (also *lysigenous*), and in many wetland species.

An interesting area of recent research concerns Radial Oxygen Loss (ROL) from roots into the rhizosphere (the immediate volume surrounding a root hair) and the consumption of oxygen by micro-organisms, or by oxidation of reduced chemicals (Colmer 2003; Visser *et al.* 2000). Advantages of ROL to the plant include such physiological considerations as an aerobic rhizosphere, facilitated mineral uptake (of, for example, NO_3^-), aerobic respiration and oxidation of iron and manganese, chemicals which are otherwise phytotoxic in their reduced state. On the other hand, ROL may result in limited growth in anaerobic soils because of reduced transport of oxygen to the growing root

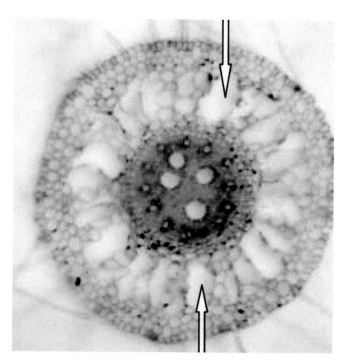

FIGURE 30. Aerenchyma (examples arrowed) in *Festuca arundinacea* induced by flooding via the plant hormone ethylene. It is thought a similar response occurs in *Agrostis stolonifera*.

tips. Not surprisingly, wetland plants differ in their response to ROL. Strong barriers in the epidermis and exodermis in roots can inhibit ROL and this can be seen in species such as *Juncus*, *Glyceria* and *Phragmites* (the principal species in the S5a and S14c communities at Britford and Castle Acre). These species have constitutive aerenchyma, and can be considered as stress-tolerant (Grime *et al.* 1990). Of greater interest for our understanding of floated communities, however, are perhaps the plants that induce barriers only in response to flooding. Recent observations, kindly communicated by Eric Visser (pers. comm.), show inducible aerenchyma in *Festuca arundinacea* (*lysigenous* form: Figure 30), triggered by the plant hormone ethylene (which builds up in cells soon after flooding) which, in turn, produces extension of shoots and rhizomes. Could a similar strategy be employed by *A. stolonifera* and other analogous species found in the high-yielding meadows managed by floating?

Why was floating abandoned at Castle Acre?

East Anglia has been described as an area in which the technique of floating was adopted at a late stage, and only ever unevenly, for a range of environmental and agrarian reasons (Wade Martins and Williamson 1994 and 1999b). The topography of the region's valleys, it has been suggested, made the construction of water meadows expensive, while the peaty nature of the valley soils, coupled with local climatic factors, ensured that the technique of floating was less effective than in the west, either as a method of forcing early grass growth or as a way of enhancing the summer hay crop. The majority of irrigated meadows, according to this view, were only created during the boom years of the Napoleonic Wars, and largely for reasons of fashion rather than as a sound agricultural investment. Most were abandoned at a relatively early date, turnips and, subsequently, oil cake – it has been suggested – providing alternative sources of winter feed (Wade Martins and Williamson 1994; 1999b; see also Williamson this volume). In this context, the Castle Acre water meadows provide an interesting comparison to Britford. Constructed by Thomas Purdey, a tenant of the agricultural innovator *par excellence*, Thomas William Coke, in or soon after 1808 (Wade Martins and Williamson 1994, 27–30), the meadows are, like those at Britford, of bedwork construction. Water was taken off the River Nar about 0.5 km upstream and caused to flow through a large carrier (about 3 m wide by 3 m deep) down on to the meadow (comprising around 30 acres (12 ha) of parallel panes) before returning to the river. The earthworks remain today, although precise interpretation of the direction of water flow is incomplete (Wade Martins and Williamson 1994, 28). Their construction evidently involved considerable effort and expense.

Whatever the truth or otherwise of the arguments concerning the general environmental unsuitability of East Anglia for floating water meadows, the valley of the Nar – a river fed in large part by chalk springs – was evidently well suited to the technique. In the late eighteenth and nineteenth centuries examples of complex and sophisticated water meadows were constructed upstream from Castle Acre at East Lexham, West Lexham, Kempstone and Mileham (Wade Martins and Williamson 1994, 23–5) but these – and those at Castle Acre itself – may in fact have been renewals, extensions or modifications of simpler, pre-existing systems. The English Nature survey of the River Nar (Cummings 1991/2) drew attention to the unusual topography of a series of meadows upstream and downstream from Castle Acre which appear to have been flood meadows. The river seems, in a number of places, to have been partially sluiced and the adjacent meadows temporarily flooded in early spring to stimulate grass by raising the temperature of the soil through the inundation of the sward with chalk-fed river water. Surveys of both the Rivers Nar and Wensum reveal countless small channels and decaying wooden structures which appear to represent simple forms of drowning, each adapted to the particular topographic circumstances of these valleys. There are grounds for believing that the Castle Acre meadows themselves incorporate parts of an earlier catchwork system.

The Castle Acre water meadow as we see it today was a grand experiment: but inherent in its grandeur may have been its demise. Growing amongst the panes of the bedworks today are the familiar echoes of floating: *A. stolonifera* on the panes and, in the ditches, *Glyceria maxima*, *Typha latifolia*, *Juncus effusus* and *Caltha palustris*. Two unusual plants, however, stand out from all the rest – the horsetails *Equisetum palustre* (on the panes) and *E. fluviatile* (in the ditches). Horsetails are considered to be poisonous to cattle. Their appearance on a water meadow designed for hay production would have been a worrying development. Horsetails can produce sporophylls and 'roots' from each node, and should an intact internode become detached from the main plant, it can float away and recolonise a new location with relative ease. Vegetative reproduction of this kind can result in very rapid colonisation of a new habitat, and once established the plant is almost impossible to eradicate. The plant's robustness largely derives from the high level of silicon in its tissue; the plants grow rapidly from rhizomes which form large clones and penetrate the ground to a depth of many metres in order to gain nutrients (Page 1997).

A recent study of *E. fluviatile* found that an experimental increase in temperature of 2.5–3 °C had a positive effect on the emergence of shoots and shoot growth, as well as increasing the maximum length of shoots (Ojala *et al.* 2001). Floating has been shown to cause soil temperatures to rise above those of neighbouring unfloated soils by between 2 °C and 3 °C, and to maintain winter temperatures at around 5 °C (Cutting *et al.* 2003). Floating may, therefore, have encouraged the growth of *E. fluviatile* which, if it colonised the Castle Acre water meadow during its flourishing years, would have presented a serious obstacle to management. The hay would have been rendered useless and removing the plant would have been difficult. Cattle grazing no doubt continued, since cattle are selective grazers, but high expectations for prolonged hay yields – the *raison d'être* of floated water meadows in the east of England – must have been severely challenged and this may ultimately have been the cause of the meadow's decline. Whether similar factors led to the demise of other meadows in similar locations in the east of England is an intriguing question.

Conclusion

The evidence from Britford suggests that regular floating over a period of more than 300 years has selected certain species over others. Research on *A. stolonifera* (the dominant species of the MG11 community, along with *Festuca rubra* and *Potentilla anserina*) appears to suggest that it is a beneficiary of the regular wet–dry cycles inherent in the floating technique. The principal reasons for this are morphological adaptations and significantly increased shoot dry weight. During floating, shallow rooting and rapid stoloniferous growth enables escape from anaerobic soils (conditions that would increase with depth): this is a plastic response to flood-induced stress. After floating, when full aerobic conditions resume, shoot dry weight increases rapidly and gives rise to the characteristically high yields observed in experiments and recorded in historical documents. The overriding anatomical adaptation appears to be the induced formation of aerenchyma in response to flooding, thus helping to maintain a constant supply of oxygen to the growing root apex. The net effect is a highly successful plant which competitively pre-empts space and dominates the vegetation community.

Further research is required in order to enhance our understanding of the ecophysiological significance of different types of aerenchyma induced by flooding. So far, it is known that there is a flux of O_2 into the soil during floating. It is also known that the permeability of roots to O_2 varies widely between species, with those less tolerant of prolonged flooding having high ROL while well-adapted wetland species have a strong barrier. The ability to induce formation of aerenchyma in species such as *Festuca arundinacea* and possibly *A. stolonifera*, coupled with rapid stoloniferous growth, is a particular advantage in floated water meadows. The planned increase in irrigation at the Harnham water meadows by the Harnham Water Meadows Trust will be an opportunity to test some of these conclusions. It will be interesting to see if *Senecio jacobea* does indeed decline as predicted, giving way to a sward more typical of MG11.

CHAPTER EIGHT

The Hydrology,
Soils and Geology of
the Wessex Water Meadows

Hadrian Cook

Introduction

Bedwork water meadows represent the zenith of a technology that was once widespread across Europe. In England, such meadows reached their most sophisticated form, and were most widely created, in Wessex – that is, the Hampshire basin and surrounding areas in Dorset, Wiltshire and Hampshire – in the course of the seventeenth century, as Joe Bettey explains in Chapter Two. By the middle of the eighteenth century there was something like 40,500 hectares of irrigated meadows in the district (Adkin 1933). From the middle of the eighteenth century, attempts were made to establish bedwork irrigation in other parts of England, and George Boswell's book *A Treatise on Watering Meadows*, published in 1779, opens with the words: 'All lands, which lie low and near the banks of rivulets, brooks and springs are capable of being watered, wherever the water is already higher than the lands, and kept within its course by the banks'. Yet, in spite of such optimism, floating with bedworks remained essentially a Wessex practice. Meadow irrigation can, it is true, successfully operate in a range of topographic and hydrological environments, but it is only in the classic chalk stream valleys of Wessex that the practice became, for several centuries, an essential ingredient of the farming regime. This chapter looks at some of the hydrological reasons why this was so: the features of water character and water supply, and of the geomorphology of the Wessex valleys, which made the meadows of this region peculiarly well suited to this form of management.

Of course, a whole range of cultural factors also played a part in Wessex's sustained supremacy in irrigation. Where innovation is followed by successful take-up, that is, sustained use for a protracted period of time, it is seldom the outcome of one factor (Dugan and Dugan 2000). The adoption of a new technology depends not only on its actual invention, but also on its promotion by the influential and wealthy – in the case of the Wessex meadows, by major landowners like the Earl of Pembroke (Cowan 2005) – and, above all, on the perceived economic benefits that make investment in novelty worthwhile. Irrigation, as we have seen, improved the quality of the grass sward, provided an 'early bite' and increased the hay crop, all of which allowed stocking densities to be raised. This was of particular importance in this region because of the benefits it brought to existing forms of sheep-corn husbandry. The downlands are well suited to cereal cultivation, particularly of barley, but the thin chalk soils (such as those of the Upton 1, Andover 1 and Andover 2 Associations (Findlay *et al.* 1984)) needed to receive continual manuring, primarily through the systematic folding of sheep on the fallows. The size of flocks was limited by the shortage of winter fodder: water meadows thus provided a way of increasing not only the output of meat and wool, but also cereal yields. In addition, long experience of the essential techniques associated with a particular innovation is also an important factor in successful diffusion. The region has a long history of water management: the construction of water mills and their leats goes back to Anglo-Saxon times; while medieval alterations to major rivers, to improve transport, culminated from the seventeenth century in the construction of canals, some of which were also used to supply water meadows.

None of these factors, however, would have mattered if the local environment had not been peculiarly well suited to irrigation. Indeed, so suitable were the Wessex valleys to bedwork floating that the technique continued to be used by local farmers even when the exploitation of the valley floors changed in the later nineteenth century, with an increasing reliance on dairy farming. In other words, while the economic and social fabric of post-medieval Wessex may have favoured the development of bedwork irrigation, environmental variables – including soil and water temperatures, water chemistry, local geology and geomorphology – were all crucial to the successful adoption, and operation, of this form of management. For in valleys such as those of the Avon, Meon, Itchen (IGS/SWA 1979) and Piddle, all in Hampshire, and the Frome and Stour in Dorset (IGS/WWA 1979), conditions were almost perfect for successful irrigation.

Hydrogeology of chalk streams

Chalk is a soft and unusually pure form of limestone dating to the later Cretaceous period (136 to 65 million years ago). It dominates the geology, and hence the landforms, across large areas of southern and eastern England, as well as being a significant feature of the surface of much of Europe, including parts of Denmark and northern France. In the Hampshire basin, which embraces the counties of Hampshire, much of Wiltshire and Dorset, and the Isle of Wight, the formation is more than 480 m thick (Chatwin 1960) and is divided into the Lower, Middle and Upper Chalk, the two latter being hard and pure forms of calcium carbonate which contain flint. The Lower Chalk, which contains no flint, is less pure (IGS/SWA 1979).

Chalk is typified by fissures which arose from post-depositional shrinkage. These are set along the bedding planes, and approximately at right angles to them. Although chalk is porous, approximately 40 per cent of the upper chalk has pores less than 1 μm (0.001 mm) in diameter, so that the storage and transmission of water is dominated by the orthogonal system of fissures, which typically lie 0.1 to 0.2 m apart and have a width of anything between 0.05 and 5 mm. The hydrogeology of such rivers as the Avon, Bourne, Test, Itchen and Meon is dominated by the Upper Chalk, although in the south of Hampshire younger beds – principally sands and clays of Tertiary age – also exert an influence on tributary streams (IGS/SWA 1979).

The chalk aquifer (aquifer literally means 'water bearer') acts, in effect, as a giant sponge. It absorbs a large proportion of the region's rainfall – only a small proportion runs off at the surface – mainly between November and April in normal years; and discharges this at springs and seepage zones throughout the year. As a result, a large proportion of the water flowing through the local rivers has first passed through the chalk. Figure 31 shows the pattern of flow for the river Nadder at Wilton in the form of a *hydrograph*: that is, a plot of the river discharge over time. It shows some muted peaks arising from rapid runoff following individual 'rainfall events', but clearly indicates that – typically for the region – the bulk of the flow is routed through groundwater.

A useful measure of the relative contribution of surface flow and groundwater discharge to the volume of water passing along a river channel is the Base Flow Index, or BFI (Marsh and Lees 2003). BFI is a dimensionless value computed by the National Water Archive at the Centre for Ecology and Hydrology, Wallingford, using gauged daily mean flow records from archived data, and may be defined for convenience as *the proportion of the river runoff that derives from stored sources*, mainly permeable rocks of the 'solid' geology, drift (superficial geological deposits), and soil materials. The potential for storing water *within permeable materials* is an index of the ability of the catchment to sustain river flow during dry weather. For example, rivers draining impervious catchments and which have minimal lake or reservoir storage will typically have a BFI in the range 0.15 to 0.35. The Nadder at Wilton has a surface catchment area of 220.6 km² and the underlying solid geology includes chalk: here, the BFI is 0.82, but some chalk streams have a BFI greater than 0.9 (Table 4).

The first five rivers in the table, noted for the extent of their bedwork meadows, all have base-flow components in excess of 80 per cent; the remaining five, in other parts of southern Britain, have lower BFIs because they have geologies with much lower permeabilities.

Figure 32 (redrawn from IGS/SWA 1979 and IGS/WWA 1979) shows the hydrogeological character of the river Avon, and of that of its various tributaries which converge at Salisbury. The solid geology of their catchments is dominated by chalk, overlain locally by Tertiary age clays and sands; these are in turn overlain by patches of plateau gravel. Within the valleys themselves, valley-fill gravel derived from the meltwaters which over-deepened the valleys during and at the end of the Ice Age are also present. These outcrop to the side

River Catchment Gauging station	County and Grid Reference	Mean Flow (cumecs)	Base Flow Index	Years of Record
Avon at Amesbury	Wiltshire SU 151413	3.49	0.90	1965–2000
Nadder at Wilton	Wiltshire SU 098308	2.90	0.82	1966–2000
Frome at East Stoke (total)	Dorset SY 866867	6.41	0.85	1965–2000
Itchen at Riverside Park	Hampshire SU 445154	5.4	0.92	1982–1999
Test at Broadlands	Hampshire SU 354189	11.01	0.94	1957–2000
Tamar at Crowford Bridge	Cornwall SX 290991	2.26	0.29	1972–2000
Beult at Stile Bridge	Kent TQ 758478	2.06	0.24	1958–2000
Adur at West Hatterell Bridge	West Sussex TQ 178197	1.07	0.25	1961–2000
Great Stour at Wye	Kent TR 049470	2.21	0.58	1962–2000
Medway at Chafford Weir	Kent TQ 517405	3.08	0.49	1960–2000

Table 4. Base Flow Indices and long-term flows for selected rivers (Marsh and Lees 2003) at selected gauging stations. The first five are chalk rivers famous for their water meadows, and the following five are a selection of other rivers, all of which have noticeably lower BFIs.

of the modern alluvium (Findlay *et al.* 1984), forming distinct terraces; Sealy (1955) identifies no less than eight in the Avon valley to the south of Salisbury, between Ringwood and Bournemouth. Gravels occur in places beneath the alluvium, forming layers more than 4 m thick to the south of Britford. Clay is also often found on the margins of the valleys, near to the valley sides, interposed between gravels and chalk. The most important thing to note in Figure 32 is the way in which the water table in the chalk tilts both down the valley, and towards the valley floor, on the east and west sides respectively. This means that, normally, the river is continuously replenished from water both seeping sideways and upwelling beneath the floodplain.

Figure 33 is generated by FLOWNET version 5.12 (van Elburg *et al.* 1989) using hydrogeological parameters published in Monkhouse and Richards (1982). It shows the 'equipotential' lines (dashed) and flowlines (dotted) for water moving in response to the (low) water table elevations from IGS/SWA (1979). Bars at the top of the diagram represent these differing elevations. Water in the chalk moves towards a broad floodplain, such as that at Britford, from both east and west (left and right of the diagram). The direction of flow is along the dotted flowlines, replenishing the river in the middle (lower) portion of the diagram, where there are no bars. From BFI values, it appears that perhaps 80 to 90 per cent of river flow is routed through the chalk and younger deposits, rather than reaching the river channel directly from surface runoff.

The hydrographic characteristics of the Wessex valleys thus made them particularly well suited to meadow irrigation, for the simple reason that the

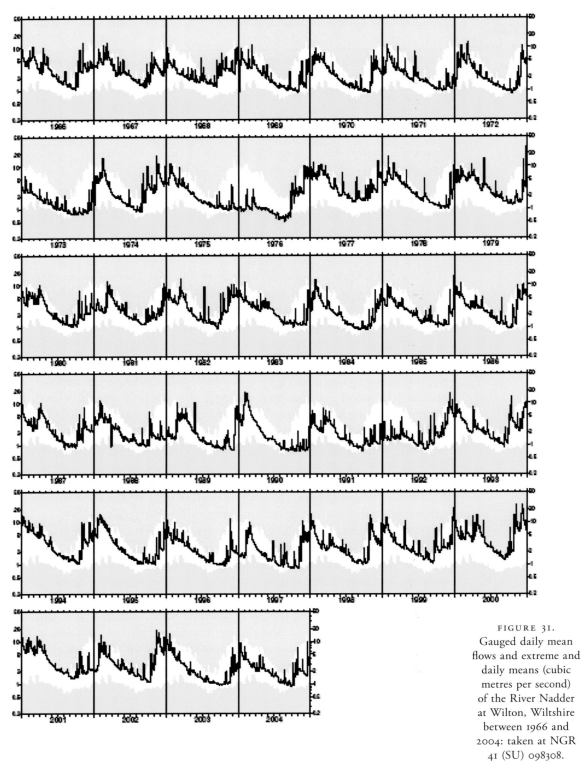

FIGURE 31.
Gauged daily mean
flows and extreme and
daily means (cubic
metres per second)
of the River Nadder
at Wilton, Wiltshire
between 1966 and
2004: taken at NGR
41 (SU) 098308.

98

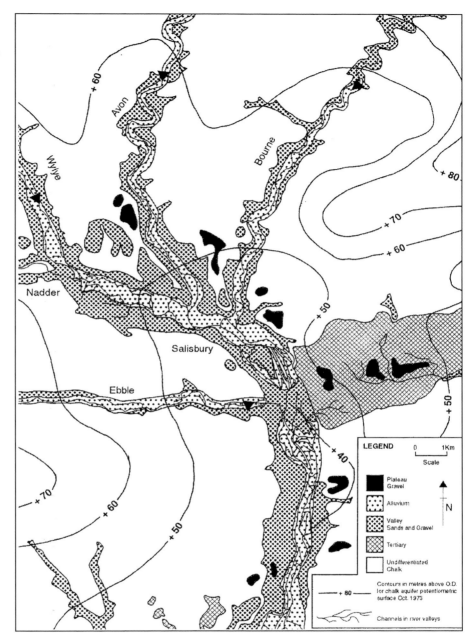

FIGURE 32.
The hydrogeology
of the area around
Salisbury, Wiltshire
(redrawn and
simplified from
IGS/SWA 1979 and
IGS/WWA 1979).
The complexity of
the channel network
reflects the extent of
human interference
in the form of river
diversions and canals,
carriages and tail
drains for water
meadows, and mill
leats.

flow of the principal streams and rivers is relatively even throughout the year
and is maintained during periods of low rainfall. Whether the main inten-
tion of the farmer was to obtain an 'early bite', or an enhanced hay crop, this
dependable water supply was a strong incentive to make the kind of heavy
investment which irrigated meadows, and especially those of the 'bedwork'
variety, demanded.

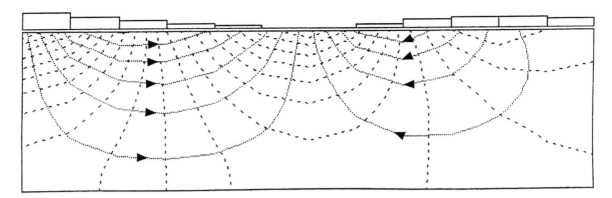

Soils and water meadows

Most of the Wessex water meadows, including those at the Britford SSSI (Cuttings *et al.* 2003), lie on soils of the Frome Association (Findlay *et al.* 1984), calcareous alluvial gleys with a silty clay loam texture. A minority occupy areas of more peaty soils, of the Gade and Adventurer's series, which have developed in shallow depressions in the valley floors, or soils of the Racton series, similar to Frome soils but with non-calcareous subsoils. Frome Association soils typically overlie deposits of gravel, with chalk and flint pebbles in a sandy or loamy matrix, at a depth of 350 to 500 mm. On the valley sides, river terrace soils are typified by soils of the Hucklesbrook Association: Hucklesbrook series soils are coarse loamy non-calcareous typical brown earths, overlying gravel at a depth of 0.8 m or less (Cutting *et al.* 2003).

Atwood (1963) noted the old saying of the 'drowners', that 'You lets the water on at the trot and off at the gallop.' This would have the effect both of providing an even depth of flow, and of maintaining the oxygen status in the irrigating water (see Cutting and Cummings, this volume, pp.77–9). The ridging-up of the alluvium to create the 'bedworks' was intended to provide such a steady and relatively rapid flow, and an even coverage, ideally to a depth of around 25 mm above the root-zone of the sward, leaving much of the aerial parts of the plants above the water level. Bedworks typically comprise ridges 6 to 10 m wide and 0.5 to 1.0 m in height (Sheail 1971): the creators of the Wessex meadows were thus fortunate in having deposits of gravel beneath the alluvial soils, which not only provided a firm base for the construction of the 'panes' but also greatly assisted drainage, although in some places this was impeded by the particular characteristics of these deposits. A loamy texture will not conduct water with anything like the efficiency of a sand-dominated matrix, and there are often clay lenses and indurated 'hardpan' bands within the gravels that would certainly impede drainage (Moon and Green 1940).

Maintaining a head of water is crucial to the successful operation of bedwork water meadows. At the point of irrigation, the water needs to be at least a metre above the level of the river, something which is achieved by taking

FIGURE 33.
Flownet for the chalk at Britford, Salisbury (generated using FLOWNET version 5.12 using data from IGS/SWA 1979, for low water tables mapped for October 1973 (low water scenario). Ordnance Survey topographic information and aquifer characteristics from Monkhouse and Richards 1982. Model width is 12 km with closed left, right and lower boundaries.

FIGURE 34.
Floodplain long-profile
along the Salisbury
Avon between
Amesbury Bridge
and Christchurch,
showing variations in
gradient. The location
of the major water
meadow systems are
indicated. (Drawn
from Ordnance Survey
1:25,000)

water along leats with a gradient less than that of the river valley itself. Moon and Green (1940) suggest that a gradient of 1:400 was required to float the meadows of the Avon 'at least along the small carriers and top gutters and between Downton and Ringwood'. Along the length of the river from weir to weir the gradient is 1:1,200, reducing considerably to the south. To maintain heads sufficient to water the meadows, therefore, main carriers needed to be relatively long. The main carrier at Lower Woodford is thus 1.2 km in length; the Avon Navigation (which supplies the Britford SSSI) is 3.7 km; the carrier opposite Charlton (on the east side of the floodplain) is 1.6 km; while one passing through Downton, on the west side of the floodplain, is no less than 4 km long. Nevertheless, the valleys of these chalk streams fall more rapidly along their length than many rivers, thus facilitating the construction of bedwork systems. In addition, their floodplains are normally wide, relative to the size of the watercourses they now contain: 'they are memorials of the time after the great Ice Age when the water-table in the chalk was higher and the volume of water much greater' (Stamp 1950, 71).

Figure 34 shows the gradient of the river floodplain between Amesbury and Christchurch, a stretch of river famous for its complex water-meadow systems. The total fall is 63 m and the mean gradient is 0.0012 (close to 1:800). The long-profile of the valley is not smooth, however, and it is noticeable how the

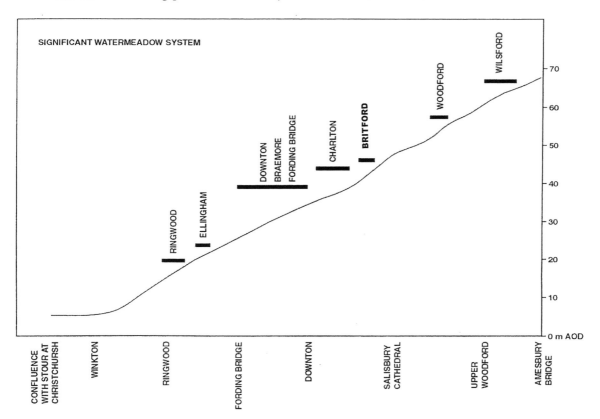

main areas of water meadow are located where this figure is approximated or exceeded. The relationship between the layout of the various meadow systems and the pattern of river channels is also worthy of note. At lower Woodford, for example, the principal carrier comes off on the west side of the river and runs approximately through the middle of the system as the river loops to the east (Cowan 1982 and 2005). The floodplain to the south of Salisbury widens to approximately 1.5 km and in the seventeenth century this presented a great opportunity for water meadow development. The system on the eastern side of the river is controlled from Wyre Hatches, where the river divides into three and feeds five main carriers (the longest being in the order of 1.5 km) which irrigated a highly planned and regimented system, covering perhaps a square kilometre, within the loop of the Avon between Britford village and Longford Castle. At this point, carriers come off the main carriers to form a herringbone pattern, but only a part of the large system at Lower Farm, Britford, remains fully operational.

At the Britford SSSI, in the area to the west, the system is fed from the relatively straight Avon Navigation (long redundant as a canal for water transport). This is taken off the main river just below Harnham Bridge, at a point approximately 2 km to the north-west of the meadows. In this area the individual meadows bear personal names (Water's, Wyatt's, Butcher's and Verne's), with each containing a more chaotic pattern than that evident elsewhere in the area (Cutting *et al.* 2003), perhaps because of the complex and fragmented pattern of land holdings here (Figure 35). Further down the valley, the relatively orthogonal system opposite and above Charlton-All-Saints is conveniently contained within a loop in the river Avon, as are the systems between Braemore and Fordingbridge.

The temperature and hydrochemistry of chalk waters

At the heart of meadow irrigation lie two apparent paradoxes. In normal circumstances farmers strive to remove excess water from damp soils in order to increase productivity. For example, reclaimed grazing marshes usually produce a better hay crop when the water table is lowered (Cook 1999). Moreover, irrigation is usually applied to dry soils, during the growing season. Meadow irrigation, in contrast, is also applied in the winter and early spring, and when the soils are at 'field capacity' or wetter, thus contradicting two apparent norms of land management.

As explained elsewhere in this volume, the benefits of winter irrigation principally arise, not from the water *per se*, but from the increased temperature it brings to the grass sward. The medieval climatic optimum of the twelfth and thirteenth centuries was followed by the 'Little Ice Age', a cooler period which ran from approximately 1500 to 1850 (Wilson *et al.* 2000, 43). This, it should be noted, coincides with the main period of water meadow construction – a point seldom if ever emphasised in discussions of meadow irrigation. Mean annual temperature over this period varied both in time, and around the globe, but for England is estimated to have been between 0.2 °C and 0.7 °C lower than in the mid twentieth century, with the effect that the growing season was shorter by as much as one or even two months. Figure 36 shows estimated prevailing temperatures during the whole year (a); in high summer (b); and, most importantly, in the winter months (c) (Lamb 1995). The benefits which winter irrigation conferred, in terms of warming the soil, protecting the sward from frost and stimulating grass growth, were thus probably more important than they would be today.

The particular characteristics of the Wessex rivers are relevant here. As already noted, they are fed mainly from water passing through the chalk. As well as ensuring that the rate of flow is remarkably even and steady throughout the year, this also means that the temperature of the water is significantly higher than if the rivers were fed primarily from surface runoff, and even in the middle of the winter the water flowing across floated meadows is usually well above the 5.5 °C necessary to stimulate grass growth. Experiments at Britford have shown that soil temperatures are appreciably greater in floated than in unfloated parts of the meadow (Cutting and Cummings 1999; Cutting *et al.* 2003). But other features of the water in these chalk streams also greatly assisted successful irrigation: the water flowing across the meadow, and infiltrating the sward, maintains its oxygen status, and this helps to prevent the anaerobic conditions which would be harmful to grass (Cook 1999).

Dissolved ions in chalk groundwater are dominated by calcium and bicarbonate (IGS/SWA 1979; IGS/WWA 1979). At source, the calcium content can be 95 to 120 mg/l and, while the levels are lower by the time the water reaches the meadows, they remain significant. The water taken from the Avon Navigation to irrigate the meadows at Britford, for example, has calcium levels of around 40 to 50 mg Ca l^{-1} (Cook *et al.* 2004): the pH measured between 1996 and 1998 ranged from 7.34 to 8.10, oxygen saturation means ranged from 85 to 92 per cent throughout the system, while the pH of the topsoil ranged from 7.39 to 7.79.

The species of nitrogen that is dominant in river water, and hence in the water used to irrigate the Wessex meadows, is nitrate (NO_3-N). This reflects the high oxygen status of the water (Environment Agency 1988). Mean determinations (November to January) for the river Avon at Salisbury, for the years 1985 to 1998, ranged from 4.9 to 6.3 mg l^{-1} NO_3-N, 0.03 to 0.05 mg l^{-1} NO_2 and 0.03 to 0.3 mg l^{-1} NH_4-N. The Wessex water meadows thus operated in a calcium-enriched and high pH environment, and it is noteworthy that many eighteenth- and nineteenth-century writers expressed a preference for alkaline waters in meadow irrigation. Marshall, for example, writing in the 1790s, emphasised the superiority of calcareous water for catchwork irrigation in Devon (1796b, 208–9), while Paxton (1840) writing of Oxfordshire, clearly appreciated the benefits to be derived from 'a spring oozing out of limestone and marl strata' when describing the construction and layout of his meadow at Bicester.

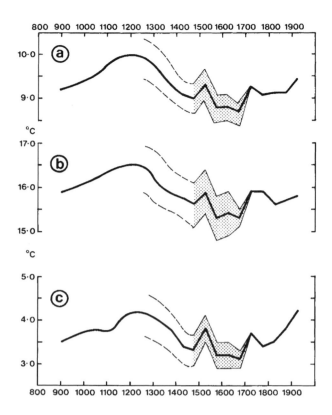

FIGURE 36. Estimated prevailing temperatures during the whole year in the period AD 900–1900 (a); high summer (b); and winter months (c). (After Lamb 1995).

All this is of particular importance because of the extent of the damage caused to grass by even short-term waterlogging. If waterlogged conditions persist for no more than forty-eight hours, all the soil oxygen is likely to be consumed by aerobic respiration within the soil. The consequence for most plants (rice and reeds excepted) is serious. The roots will cease to respire, and the microbial population switches from aerobic to anaerobic respiration pathways. Under field conditions, individual peds (soil aggregates) may remain wet, causing their interiors to become anaerobic (White 1997) due to the slow diffusion of oxygen in soil waters. Be they organic or mineral in character, when soils become saturated with relatively immobile water, oxygen cannot diffuse through the pores and anaerobic (or reducing) conditions rapidly result. Oxygen diffusion rates in wet soils are some 10,000 times slower than for drained soils so, without replenishment, the soil biomass soon becomes oxygen deficient.

When a soil contains free oxygen, either gaseous in the soil air or dissolved in the soil water, aerobic respiration (essential to the root system), mineralisation, and the conversion of ammonium to nitrate and oxidation of sulphide to sulphate can all occur. Avoiding anaerobic conditions is therefore the key to avoiding phytotoxic compounds, as well as the loss of nitrate to gaseous N_2O and N_2. Nitrogen is also released through protein decomposition as NH_4^+, which can be the dominant form of nitrogen in grassland soils. Furthermore, a high soil-water pH makes denitrification unlikely, because the redox thresholds of nitrogen compounds fall with increasing pH (Wild 1988). Organic compounds may also result from anaerobic conditions through fermentation. One group of substances, termed Volatile Fatty Acids, may be created through carbohydrate fermentation arising from crop residues (White 1997). These are toxic to plants, especially young seedlings, but due to their low molecular weights are lost to the atmosphere as CO_2, CH_4 and H_2 through the action of anaerobes.

Increasing anaerobic respiration therefore has a number of serious consequences beyond the cessation of ordinary respiration, including the evolution of organic and inorganic phytotoxic substances (such as ethylene) and the loss of dissolved nitrate in the soil to the gaseous phase through microbial denitrification. A saturated soil will experience a series of reduction reactions as anaerobic processes 'switch in' where there is decomposable organic matter. The oxidation-reduction, or *redox potential*, of water is an important variable in controlling chemical reactions in aquatic, soil and groundwaters. The redox potential is a measure of electron availability in a solution and quantifies the degree of electrochemical reduction in soils. Reduction is when oxygen is given up, hydrogen is gained (hydrogenation), and electron gain occurs; oxidation is the reverse of this process. Oxidation-reduction potential is conveniently measured in millivolts (mV) on the 'Eh scale' using an inert platinum electrode. Where free, dissolved oxygen is present ('aerobic conditions'), the range is generally between +400 and +700 mV; in reducing ('anaerobic') conditions, the range tends to be between +400 and −400 mV. In a waterlogged soil, where organic substrates are oxidised and the redox potential falls (i.e., moves towards or into the negative range), electron gain occurs. The secret of successful irrigation is to maintain levels of +400 mV or more in the topsoil and thereby avoid reduction (Cutting *et al.* 2003).

Eh is both temperature and pH dependent, but pH is the strongest influence of the two: a reaction which reduces nitrate will fall in Eh potential by 59 mV per unit increase of pH, although the chemical complexity of field soils means that the reactions are only partially applicable in this context, and indeed a wide range of the Eh_7 thresholds for reduction are given by different authorities (Correl and Weller 1989).

The moving water which enters an irrigated meadow in Wessex in the winter months, when irrigation for the 'early bite' takes place, has a high pH and high levels of oxygen saturation. It is also, as already noted, commonly at a temperature of 5.5 °C or more. For water that is in equilibrium with the atmosphere – i.e., saturated – and at standard pressure, oxygen content at 5 °C is 12.4 mg/l, falling in a non-linear fashion to 11 mg/l at 10 °C and 8.7 mg/l at 20 °C (Moss 1988). Cooler water can dissolve and hence transport more oxygen.

Water meadow irrigation is thus analogous with hydroponic propagation. Because the surface layer of the system is buffered against acidity, and constantly dissolving oxygen from the air, adverse soil conditions (particularly anaerobic conditions) are avoided. Moreover, nitrogen is predominantly in the form of nitrate, and unlikely to become lost to the system as N_2O. Indeed, the analogy with hydroponics systems is so strong that the author has observed green grass shoots growing where the sward has been cut (for the purposes of measuring productivity) *before* the water was removed.

Conclusion

Looked at from the point of view of the hydrologist, it is thus easy to see why water meadows, and especially those of the bedwork variety, were more successful in Wessex than in any other region of Britain. The gradient, and the width, of the principal river valleys, together with the character of their floodplain deposits, made both the construction and the operation of bedworks extremely cost-effective, while the hydrogeography of the rivers ensured a dependable flow of water which, in the winter months, had a temperature which was consistently higher than that of the air. Lastly, the chemistry of the river water helped prevent the build-up of harmful anaerobic conditions, while serving, in a variety of complex ways, to improve the quality of the grass sward. Whatever the economic, agrarian and social factors which encouraged the development and widespread adoption of bedwork floating in Wessex, it was these environmental characteristics which really made the technique so appealing to farmers and landowners in the seventeenth and eighteenth centuries.

CHAPTER NINE

The Management of Water Meadows: Four Hundred Years of Intensive Integrated Agriculture

Kathy Stearne

Introduction

Water meadows formed an important part of the agrarian economy of southern England for over 400 years. Meadow irrigation was certainly practised in some form in the Middle Ages (Cook *et al.* 2003) but it was in the period between the sixteenth and the nineteenth century that valley bottoms in southern England were most intensively exploited in this way. Much has been written about the economics of water meadows, their development and diffusion. Surprisingly little attention, however, has been paid to the details of their operation, or how this changed over the centuries.

Forms of management before the late nineteenth century

The earliest text to provide any real details of the annual management of water meadows is that written by Rowland Vaughan at the end of the sixteenth century (Vaughan 1610). Vaughan experimented with surface irrigation to increase grass production in the Golden Valley in Herefordshire between about 1580 and 1606. He describes the construction of his water-meadow systems and, in some detail, the increases in productivity which they brought about, in terms both of the early grazing they provided and the increased summer hay crop. Vaughan was, however, prone to self-aggrandisement and his claims need to be treated with a measure of caution. More importantly, in the present context, much of his detail is vague; for example, he does not give the duration of the irrigation periods in winter, nor clear information about which animals were pastured on the water meadows at different times of the year. He thus states that the meadows were grazed early in the spring, but whether by sheep or cattle remains unclear. In spite of these problems it is, however, possible to reconstruct in broad detail the management system employed by Vaughan from a close reading of his text (Table 5 and Figure 37).

By the end of the seventeenth century water meadows were well established

Season	Operation	Quotation from Vaughan 1610
Winter	Irrigate with 'muddy' water	'as it riseth muddy, imploy it on your grounds during the winter season' – p. 96 'drowne the frozen grounds when the snow begins to fall: so shall you release the grasse being bound and spend the snowe' – p. 119
Spring March /April	Drain Graze sheep or cattle	'be sure in the begininge of March to cleare your ground of cold water; and keepe it dry' – p. 97 'that all kinde of cattle, (sheepe especially), may have sufficient to sustain them' – p. 120 'I can graze my mowing meades until the first of Maie' – p. 127
May	Irrigate for a short period	'In May … You need not drowne it until a day, two or three before you mow' – p. 98
June	Hay harvest – one or two cuts	'you shall finde this manner of drowning in the morning of your mowing so profitable and good that commonly you gaine ten or twelve days advantage in growing' – p. 99 'drowning before mowing, will make good a second mowing' – p. 100
Summer	Irrigate overnight Graze cattle for two weeks at a time	'You may suffer colde water in the heate of the summer to cool your grounds' – p. 97 'In Summer drown it moderately' – p. 120 'Not drowne in the heat of the day, but in the night sunne to sunne' – p. 137 'Cattle grazing for two weeks' – p. 127
Autumn	Irrigate with muddy water	'In flood times see you suffer not your floud water by neglect to pass away into the brook … but by your sluice command it to your grounds; … so long as it appears muddy' – p. 96

Table 5. Water meadow management in c. 1600, after Rowland Vaughan

in the chalk river valleys of the south of England, and since that time the story of water meadows has been intrinsically linked with the local system of sheep-corn farming (Bowie 1987). Sheep were grazed on the meadows during the day for a full month in the early spring before the grass was sufficiently grown for them to be moved elsewhere. The water had to be applied in the winter with care, and a skilled drowner could have the grass ready for grazing a fortnight before a less skilled one. As the old Wiltshire saying put it:

'The Drowner is half of the meadow,
The Shepherd is half of the flock'.

The sheep were only grazed on the water meadows during the day, between around 10 a.m. and 4 p.m. At night the flock was folded on the arable land, where the dung and urine manured the thin chalk soils characteristic of the area. This considerably increased the yields of barley, in particular; but the disadvantage of the system was that the sheep were obliged to walk anything up to two miles (3.2 km) a day and this caused high mortality rates in lambs (Davis 1811) – although against this there was the advantage that sheep were not left for long periods on damp meadows, where they might contract ailments such as foot rot.

Water meadows gave a sustainable increase in production of cereals and

*The Management
of Water Meadows:
Four Hundred
Years of Intensive
Integrated
Agriculture*

mutton within sheep-corn systems: as a rule of thumb one acre (0.4 ha) of water-meadow grass in the spring fed 500 couples (that is, ewes and lambs), which in turn manured one acre of arable land (Kerridge 1967). So important was the production of manure that at times of high corn prices, as, for example, during the early nineteenth century, the primary use of the sheep was as dung transfer machines. In more normal conditions, when grain prices fell back, the farmer had to balance which was the more profitable – mutton and wool, or corn – taking every account of current weather and ground conditions.

At the end of the eighteenth century George Boswell (1779) wrote about all aspects of water meadows and their management. He gave very detailed information about the system, its advantages and its problems. He stressed that the hay crop was a very important product and emphasised the way in which watering improved both its reliability and quantity. He also – unlike Vaughan – provided details of the kinds of maintenance and repair work which the meadows required. After the sheep had grazed off the early 'bite' in the spring and been moved on to other pastures, the meadows were rolled, and general maintenance work was carried out, before a further spell of irrigation, followed by hay cropping. They were grazed again, this time by cattle, at the end of the summer, and this was followed by a further period of maintenance, to ensure that the meadows would be ready to accept water from the first rains of autumn: the 'thick and good' water which served to fertilise them. Boswell also provided details of the duration of irrigation, and of how this differed between summer and winter. He thus provides, for the first time, a detailed description of the water-meadow year (Table 6 and Figure 38).

He described a whole range of advantages which he believed irrigation brought. Firstly, the water manured the land, and in this context he discussed the value of the runoff from towns, farms, arable land and chalk hills, noting that, due to the accumulation of nutrients brought to the meadows by the water, their productivity increased year on year. Boswell also emphasised the consistency and predictability of the grass growth on watered meadows, even in dry summers, noting that sometimes a second crop of hay could be taken. The early grass was of particular importance, providing feed for young lambs in February, March and April. He claimed that the combined effect of early grass and an increased hay crop in some cases raised stocking rates on a farm ten-fold. Lastly, he suggested, maintenance of the meadows provided work for the 'industrious poor'. The kind of management described by Boswell lasted for over two hundred years, from the mid seventeenth to the mid nineteenth century. In 1800 Davis reported that there were between 15,000 and 20,000 acres (c. 6,000 to 8,000 ha) of water meadows in Wiltshire alone (Davis 1811). Meadows were expensive to maintain, in terms of both materials and labour: shepherds, drowners, carpenters, carters, local labourers and itinerant workmen (to cut the hay), and cowherds were all involved with the Wessex meadows at various points in the year. The produce of the meadows was not

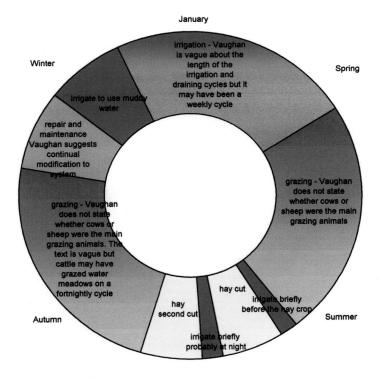

January

Winter

Spring

irrigation - Vaughan is vague about the length of the irrigation and draining cycles but it may have been a weekly cycle

irrigate to use muddy water

repair and maintenance Vaughan suggests continual modification to system

grazing - Vaughan does not state whether cows or sheep were the main grazing animals

grazing - Vaughan does not state whether cows or sheep were the main grazing animals. The text is vague but cattle may have grazed water meadows on a fortnightly cycle

hay cut

irrigate briefly before the hay crop

hay second cut

irrigate briefly probably at night

Autumn

Summer

FIGURE 37.
The annual cycle of water meadow management implied by Rowland Vaughan's *Most Approved and Long Experienced Water Workes* of 1610.

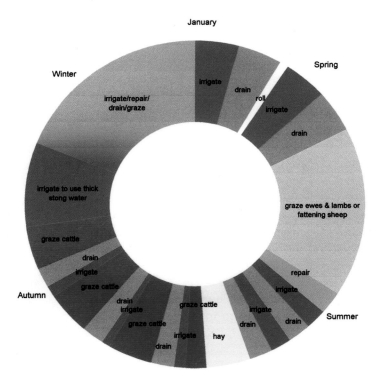

January

Winter

Spring

irrigate

drain

roll

irrigate

drain

irrigate/repair/drain/graze

graze ewes & lambs or fattening sheep

irrigate to use thick stong water

graze cattle

drain

irrigate

graze cattle

repair

irrigate

drain irrigate

graze cattle

graze cattle

irrigate

drain

hay

irrigate

drain

drain

Autumn

Summer

FIGURE 38.
The annual cycle of water meadow management described in George Boswell's *A Treatise on Watering Meadows* of 1779.

Date	Operation	Quotation from Boswell
January	Irrigate and drain	'When the days grow longer and begin to be warmer it must not be suffered
February	Irrigate and drain	to remain too long; that is after Candlemas [2 Feb] a fortnight is full long enough and the next turn [irrigation period] a week' – p. 97
March	Roll Irrigate and drain Graze with ewes and lambs	'Rolling meadows in spring … is an excellent method [which] contributes much to the grasses being cut close to the surface when mown' – p. 105 'There will be grass long enough to take ewes and lambs, or fattening sheep to finish them for the butcher; … they may be permitted to feed [on the water meadows] till the beginning of May' – p. 93
April	Irrigate and drain Graze with ewes, lambs and young cattle Carry out repairs	'Spring feed should never be done by other than sheep or calves' – p. 99
May	Irrigate for a few days	'Alternately watering and draining, lessening the time water remains upon
June	Shut up for hay	it, as the weather grows warmer, and in five, six or seven weeks, the grass
July	Cut for hay	will be fit to be mown for hay and produce from one to two tons or even more an acre' – p. 93
August and September	Irrigate and drain on a daily cycle Graze with cows	'After math … is of vast advantage to milsh beasts. The butter made from it is excellent' – p. 104
October	Irrigate with the heavy rains of October	'… these waters are always very thick and rich being the washings of all the country' – p. 99
November	Carry out repairs	'Every two or three days the workmen go and examine the trenches … water can be kept on the ground a month or six weeks' – p. 97
December	Carry out repairs Irrigate and drain	'A fortnight or three weeks is long enough for water to remain on the ground' – p. 97

Table 6. Water meadow management in 1779, according to George Boswell

only considerable, but also varied, including not only dung for the arable land but also lamb, mutton and hay, as well as (in some circumstances) milk and other dairy produce. If market prices dropped for one product, the income from the others still justified the work and expense involved: the system had a degree of flexibility, and could be adapted to changing conditions.

Indeed, we should not assume that the sheep-corn systems of the southern chalklands were stable and unchanging over time. From the beginning of the nineteenth century, for example, farmers in the district were experimenting with improved breeds of sheep. The Old Wiltshire Horn breed, though a perfect 'dung transfer machine', gave 'indifferent fleeces and worse carcasses'. It was steadily replaced on Wessex farms by Hampshire Down sheep; the Old Wiltshire Horn was nearly extinct by 1830 (Atwood 1963). South Downs sheep were also increasingly kept on the meadows, but whatever the breed, the use of the spring flocks for folding the arable could cause severe losses of both ewes and lambs during lambing. From a flock of 700 sheep, only 500 to 550 lambs were expected to survive. Caird, writing in the middle of the nineteenth century, thought that this high mortality rate was due to the daily driving of the flock to and from the water meadows, a distance of perhaps a mile or more, combined with the close and crowded conditions of the sheepfold at night (Caird 1852, 60–4). While corn prices were high, such losses from the flock could be tolerated, but when corn prices declined, as in the period immediately after the Napoleonic Wars, they ceased to be acceptable and the meadows were used more for rearing lambs than for folding.

Yet it is important to emphasise, as other contributors to this volume have done, that in areas outside the Wessex chalklands water meadows might be managed in other ways, and valued for other kinds of produce. In the Wey Valley on the Hampshire/Sussex/Surrey borders, for example, it appears that hay was a particularly important commodity from the early nineteenth century until the 1920s, and the meadows here were managed accordingly. The management cycle in Table 7 and Figure 39 has been reconstructed from the diaries of James Simmons, a mill owner at Bramshot near Liphook, which span the years from 1837 to 1842 (Bowles 1988, 9).

At the beginning of the nineteenth century hay was sold to the coaching hostels on the London–Portsmouth road. There was some decline in demand with the advent of the railways in the middle decades of the century, but the Army had a large presence in the area and they constituted a major market for hay, for the army horses, right through until the end of the First World War.

Season	Operation	Quotation from Simmon's Diary
Winter/Spring	Repair Irrigate	'Carrying in earth to fill up the ditches in Sturt Meadow' – 9 December 1839 'Mr J. Lucas has turned the water out of the stream over his meadow and has done it within the last four or eight months several times' – 25 April 1842
Early Spring (March /April)	Repair Drain	'I have had … the stream in long meadow cleared out and some shrubs put in at the turns to prevent its wearing' – 21 March 1852 'Walking across a meadow where a man was draining' – 31 March 1838
Late Spring April/ May	Graze with cattle?	[Cattle may have grazed the water meadows; in the Wey area there is no documentary evidence that sheep grazed the river meadows at this time.]
Summer June / July	Cut hay	'Mowed my clover and part of Lion Meadow' – 25 June 1838 'This day took up two loads of hay from new mill meadow' – 1 July 1837 'Finished haying yesterday' – 8 July 1837 'Finished haying!' – 11 July 1838
Summer August	Short periods of irrigation?	[It was the practice in other river valleys to irrigate for short periods of time (1 day) over the summer to stimulate grass growth.]
Late summer September	Cut second hay crop	'Finished the late haying' – 22 September 1838 'Harvesting the second cut of hay at the New Mill' – 23 September 1837
Autumn	Graze with cattle or horses?	[Cattle or horses may have been allowed to graze in the autumn and winter.]
Autumn/Winter	Repair Irrigate and drain	[In the absence of direct evidence I assume that the Wey water meadows were irrigated on rotation for about a week at a time from October to March.]

Table 7. Water meadow management in the Wey Valley c. 1840, after James Simmons.

*The Management
of Water Meadows:
Four Hundred
Years of Intensive
Integrated
Agriculture*

Meadow management in the late nineteenth and twentieth centuries

The flexibility of water meadows was such that they continued to flourish in the southern counties right through the nineteenth century. Indeed, Professor Wrightson, the Principal of the Agricultural College at Downton, in the Avon Valley, was still advocating and experimenting with watering as late as 1880. In its details his system of management was slightly different from that described by Boswell in the 1770s, but the principles are much the same (Forsyth 1875) (Figure 40). Wrightson attempted to develop the productivity of water meadows further, and provided details of grass seed mixes to improve the quality of the sward. By this time, however, radical changes were beginning to affect the agriculture of southern England. As described elsewhere in this book, a major agricultural recession began in the late 1870s. The cereal harvest of 1879 failed and there was a severe outbreak of liver fluke (Bowie 1987). After 1879 grain prices fell dramatically: wet and cold seasons ruined one harvest after another, but more significant was the arrival of cheap grain imported from North America (Perry 1974). Between 1871 and 1906 farm rents on the Radnor Estate near Salisbury fell by up to 50 per cent (Stearne 2004). Moreover, the kinds of improvements in transport which allowed the large-scale import of grain from the New World also ensured that many of the other things that meadows had traditionally supplied could now be brought in from abroad more cheaply than they could be produced in England. In particular, the arrival of refrigerated lamb from Australia and New Zealand began to undermine meat prices. At the same time, fertilisers were now being imported from abroad or manufactured at home on a large scale, so the dung from the sheep was no longer needed to manure the arable land in the same quantities as before. All this meant lower prices for almost all of the things that the water meadows had traditionally produced.

Some meadows, however, found a new use, as local farmers turned from an overwhelming reliance on sheep-corn husbandry to dairy production, an aspect of the agricultural economy that remained buoyant due to the demand for fresh milk, butter and cheese from the expanding towns and industrial areas, and the ease of transport now provided by an extensive rail network. Moreover, as the recession continued landlords found it increasingly difficult to keep existing tenants or find new ones. Some tried to deal with this situation by renewing the buildings and other infrastructure on holdings, including the sluices and channels of water meadows. The water meadows thus, to varying extents, weathered the initial storm of the recession.

Very little has been written about management of water meadows in the twentieth century, in spite of the fact that irrigation continued at many places up until the 1960s, and in a few cases still goes on. Forms of management were, however, evidently rather different from those described by Boswell in the 1770s, or even by Wrightson in 1880. By the 1930s, dairy cattle were the most important animals, grazed on the meadows from March through to November (Street 1932; 1946). Irrigation had a particular value in extending

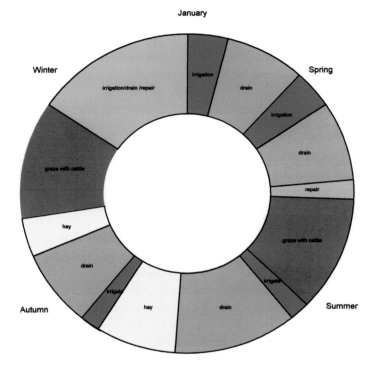

FIGURE 39.
The annual cycle
of water meadow
management in the
Wey valley, Surrey, in
c. 1840, reconstructed
from James Simmons's
diary.

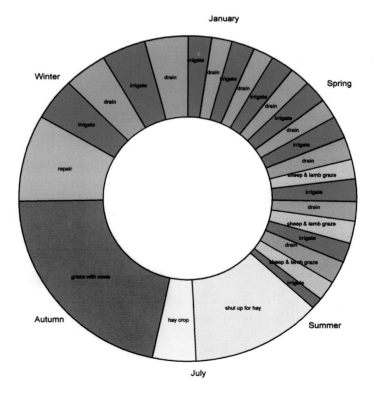

FIGURE 40.
Water meadow
management in
the late nineteenth
century, as described
in the *Encyclopaedia
Britannica* for 1880.

the growing season, for the earlier in the year and the later in the autumn the cows grazed fresh grass, the better and higher the milk yield. But the new use of the meadows caused its own problems. Cattle are heavier animals than sheep and in the spring, especially on soft peat meadows, they can cause a lot of damage to the ground. This meant that greater maintenance was required in order to keep the water meadows – with their complex but vulnerable systems of channels and drains – in good working order. The farmers said that cattle had 'five feet' – they did as much damage tearing the grass up with their mouths as they did with their feet. Martin Vining, a farmer whose father farmed land in the Test Valley near Stockbridge, remembered that grass was cut in February and March, with a sharp scythe, and carried in nets to feed cattle in the barns, in order to avoid poaching the ground. But the early season increases and the generally more reliable milk yields obtained from cows grazed on water meadows meant that farmers received a premium for their milk, making the extra work and expense acceptable.

The timing and duration of irrigation varied during the course of the twentieth century, depending on the availability of water and the extent to which it had to be shared with neighbouring farms. On the river Wylye, to judge from the various references in A.G. Street's books of 1932 and 1946, farmers only irrigated for a few days a week, and not at all in the summer – rather different from the traditional Wessex practice of fortnightly irrigation through the winter and overnight irrigation on occasions during the summer months

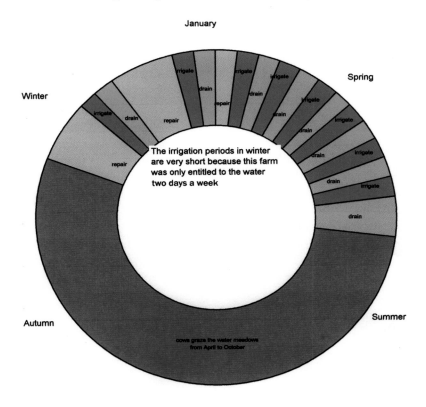

FIGURE 41.
Water meadow management in the 1930s on the River Wylye near Wilton, reconstructed from *Round the Year on the Farm* by A.G. Street (1946).

115

Month	Operation	Quotation
January	Repair Irrigate and drain	'... the drowner will be getting near the finish of his ditch cleaning. When he began to he cleaned out the carriages. Since Christmas he will have been busy cleaning out the drawing, which are the ditches that draw the water from the meadows' – p. 49
	Irrigate for two days only	'... By now the drowner will have cleaned out both carriages and drawings and will be busy placing and stopping. He turns the water into a meadow ... So he can go 'placing' and 'stopping' on only two days a week' – p. 50
February	Irrigate	'The drowning and hedging are finished' – p. 58
Late March	Drain	'Down goes the drowner to the meadows, where he shuts all the hatches that permit water on to them and opens all the hatches that drain water from them. This done, he looks around the meadows and notices with pride the rich carpeting of grass, the result of his winter work' – p. 69
April	Graze with cows	'After breakfast, the cows are driven down the hill to the water meadows ... That night the dairyman puts out a smaller allowance of hay for his cows, because he knows that once they have tasted fresh green grass they will not eat as much dry hay' – p. 72
June	Graze with cows	'Cattle sometimes fall into ditches in the meadows and cannot get out. If left too long they will die' – p. 90
November	Begin maintenance and irrigation	'On farms that have irrigated water meadows the drowner must now get busy. He is responsible for the care and cleaning of the irrigation ditches that carry the water from the main river to the meadows. During the summer the cows will have trodden in many of these and all of them will be choked up with weeds and rushes' – p. 30
December	Irrigate	'... there's six weeks' water in the meads afore Christmas' – p. 31

Table 8. Water meadow management in the 1930s on the River Wylye near Wilton (from *Round the Year on the Farm* by A.G. Street (1946)).

(Table 8 and Figure 41). Nevertheless, in the twentieth century the costs of maintaining and operating the meadows were as great as, if not greater than, in previous centuries, and farmers were now reliant on just one main product to pay for them: fresh milk, transported via the railways to towns.

Although irrigation thus survived the early twentieth-century periods of recession, the middle decades of the century saw a decline in intensive management, and many meadows were left derelict. There were a number of reasons for this: increased labour costs; a decline in milk prices; the difficulties of mechanising maintenance work; increased grass yields on other pastures; and changing social perceptions. Today there are only a few water meadows still operating in traditional ways in southern England.

Modern management: challenges and possibilities

Does this matter? Why should we be concerned with water meadow management today? There are a number of reasons, both cultural and environmental. Firstly, water meadows are an important part of our cultural heritage: John

*The Management
of Water Meadows:
Four Hundred
Years of Intensive
Integrated
Agriculture*

Constable painted them and Thomas Hardy (1878) wrote about them. They are a living historical document, a physical testament to the way life was lived in the southern river valleys of England between c. 1600 and the 1950s. Moreover, as Christopher Taylor explains in Chapter Three, they have a distinctive archaeology which includes not only the 'ridge and furrow' corrugations of the bedworks but also built structures, such as sluices and aqueducts, which demonstrate changing styles of design and the changing availability of materials over a period of 300 years or more. Some of these features, such as complex winding mechanisms, are the product of numerous manufacturers, both local and regional – Wallis and Stevenson of Andover, the 'Dutch' foundry at Warminster, or Galpin of Dorchester – a legacy of industrial archaeology the importance of which is often overlooked (Stearne *et al.* 2002).

Yet this cultural significance carries with it an inherent contradiction. On the one hand, as an important part of our historical environment, with an important educational role to play, there is an understandable desire to restore some water meadows to full working order. But to do this would, inevitably, involve some destruction of their current physical structure. Bedworks would have to be relaid, drains recut, sluices replaced. Interestingly, there are similar issues relating to water meadows as an important environmental resource. The characteristic sward created by the repeated cycle of irrigation, drainage and grazing on the meadows is relatively species-poor, although moderately interesting and now becoming rare in many districts: that classified by the NVC (National Vegetation Classification) scheme as type MG11 (*Festuca rubra – Agrostis stolonifera – Potentilla anserine*: Rodwell 1993). Richer and more diverse communities are usually found in the carriers and the ditches draining the meadows, including S5a, *Glyceria* swamp; MG8, *Cynosurus cristatus – Caltha palustris*; and MG9a, *Holcus lanatus – Deschampsia cespitosa*. Yet the hydrology and topography of derelict, non-operational water meadows has also served to produce an interesting and diverse flora and fauna. Many rare species live on old meadow sites, such as otters and water voles; many others find shelter or sustenance there. Numerous birds, such as the snipe, lapwing, kingfisher and redshank, feed and nest there, together with a host of invertebrates. Indeed, it is largely because of the extent of redundant meadows that the lower Itchen is now home to about 25 per cent of the world population of southern damselfly, an endangered species. Many of the chalk river valleys of southern England are designated Sites of Special Scientific Interest (SSSI), in large measure because of their water meadows.

At present, a relatively small number of water meadows remain in operation. A few private owners still manage their meadows in the traditional way, and Twyford Parish Council own and irrigate an example on the River Itchen. But because of their ecological and, to some extent, cultural importance, many *derelict* meadows are also maintained by conservation bodies for diverse reasons, and are managed in a diversity of ways. At Salisbury, for example, the Harnham Water Meadows Trust maintain the meadows as open grassland, the setting for the Cathedral as so famously painted by Constable (voted the

best view in Britain by readers of *Country Life* magazine in 2002). Elsewhere derelict meadows are maintained in a rougher state, sometimes even as re-generating woodland (Figure 42), and provide mainly environmental benefits, such as suitable habitats for over-wintering flocks of migrant birds. Titchfield Haven Nature Reserve, for example, located on old water meadows in the lower Meon Valley, is an internationally important wetland site designated as both a candidate Special Area of Conservation (cSAC) and a RAMSAR.* Another nature reserve on the lower reaches of the River Test is important for breeding birds, notably snipe, lapwing and redshank, while the ditches of old water meadows at the Itchen Valley Country Park provide habitats for a wide range of invertebrates, and the meadows themselves for a range of wetland flora. Management here includes measures to restore water vole habitats and encourage otters, which were re-released into the area in the 1990s.

Unwatered meadows thus often have an ecological importance; and at the same time, irrigation is no longer a serious economic proposition for most

FIGURE 42.
Derelict water meadows, colonised by alder carr, on the river Wey at Passfield in Hampshire.

* **RAMSAR** International wetland designation. Ramsar sites are protected wildfowl habitats which are designated in accordance with a convention signed at Ramsar, Iran, in 1971.

CHAPTER TEN

Water Meadow Management
Today: The Practitioner's View

Peter Martin and Kathy Stearne

Regular irrigation of water meadows on a large scale has all but disappeared from the English countryside, but fortunately there are a few individuals who have kept the practice alive, and most of these work in the valley of the river Avon near Salisbury. The Britford water meadows (SU 165275), farmed by Peter Martin and his neighbour, are one of the best examples of a working water meadow, with a continuous management history going back perhaps 300 years (Figure 45). Here the 'traditional' farming pattern of large landed estates and tenant farms has survived to the present, and over the centuries successive leases have required that the valley water meadows should be regularly irrigated. In the past the meadows were occupied by a number of different farmers, but they worked in close cooperation: the individual drowners and farmers depended on one another for water, and strict practice grew up over the centuries concerning when each farm was allowed to irrigate, usually in rotation, and who was responsible for repairing the various hatches. The farmers relied on each other to carry out the appropriate work at the right time, and if any one farmer reneged on his part of the work it had a knock-on effect on other users of the meadows. In 1963, when most of the Jervoise Estate was sold, a new water rights agreement had to be drawn up, by lawyers, to take account of the existing practice in the immediate area, and also the rights of farmers lying downstream.

Peter Martin was brought up at Odstock, on the River Ebble (a tributary of the river Avon), and learnt the craft of drowner from his father. He remembers a time when irrigated water meadows existed throughout the Avon Valley and stretched up the Ebble Valley for about 13 miles (21 km), although there was never enough water to irrigate all of these in the summer. He has seen dramatic changes in farming over the years. He took over the management of the farm from his father in 1957; this property included water meadows in the Ebble valley (though the water supply to these was limited) and the tenancy of 28 ha (70 acres) of water meadows at Lower Britford on the Avon. There were two full-time drowners then, elderly men who retired soon after Peter took over the farm.

At that time there were five farms irrigating meadows at Britford. The water

FIGURE 45.
Peter Martin
(foreground) opening
a hatch at the Britford
water meadows.

was controlled from the Hatch House, about a mile upstream. Peter remembers that it was best to keep on the right side of the River Keeper at the hatch house, or you might not get the water at the time you wanted it! In 1963, when the estate was sold, the land in the area was split into a number of different ownerships, and at this point irrigation stopped on the meadows lying immediately upstream.

For the first half of the twentieth century, and probably for much of their history, the Lower Britford meadows were managed as part of a dairy enterprise, selling milk into Salisbury. Haymaking on the meadows ceased after 1947 when a previous tenant, Berty Marks, left. Peter clearly remembers his first year as farm manager, when the farm supported an intensive dairy herd. The cattle, between 56 and 60 Ayrshires, were taken down to the meadows in the last week of February or first week of March, and grazed them on rotation until October or November, depending on the weather. As noted

FIGURE 46.
Part of the Britford
meadows being floated.

FIGURE 47.
Adjusting the flow of
water, using stops, on
the Britford meadows.

was particularly long, hard and cold, and the grass started to suffer after a while and the water had to be pulled off.

At the end of February the meadows are drained down and the stock are put on some time in late March, depending on the weather. This 'first bite' of grass is produced months earlier than it could be without irrigation. From March onwards, the meadows are sporadically 'floated' following grazing for periods of between three to seven days, depending on the availability of water, although this procedure is, for obvious reasons, unnecessary during periods of wet weather. Sporadic summer watering is undertaken to 'freshen up' the meadow, and also to help prevent selective grazing around 'dung pats'.

Management of potentially problematic 'weedy' herbs is important; nettles, thistles, docks, tussock grass and rush are a persistent problem on water meadows. They are often mowed from about mid July when the creeping thistle is in bud and before the summer thistle seeds. Trimming is best done when the cattle are grazing the meadows, as they will eat the wilted trimmings, and is sometimes repeated later in the year.

The meadows of Lower Britford have been operated almost continually for more than three centuries. They clearly demonstrate that without the use of fertilisers or agro-chemicals it is possible to maintain a high-quality, high-yielding grass sward. But water meadows are highly labour-intensive and, in modern circumstances, uneconomic: only the external support provided by agri-environment schemes makes it possible to sustain this important part of

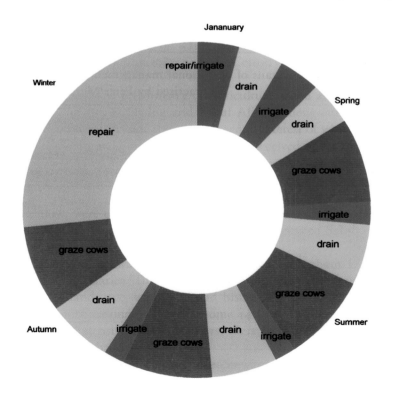

FIGURE 48.
The annual cycle of management currently followed on the Britford meadows.

Experience with both the ESA schemes and CSS has shown that even the most carefully thought-out conservation schemes can lead to disagreement and conflict. For example, a major policy objective in the Test and Avon ESA is to manage grassland in such a way as to provide a habitat suitable for breeding waders. This entails the maintenance of open wet grassland with a 'tussocky' sward, grazed lightly by cattle in the spring, for areas of standing water with muddy margins are beneficial to the birds. Meadows like this, however, look rather different from the neatly cropped, sheep-grazed grassland of the traditionally managed water meadows of the kind that those interested in historical landscape conservation might wish to see. There is a school of thought which asserts that floated water meadows were ecologically beneficial but, as we have seen (above, p.117), this does not mean they were particularly diverse in botanical terms. Many derelict water meadows have, in fact, developed a particularly diverse and interesting flora since they stopped being floated. To reinstate floating could destroy a unique ecosystem which may include invertebrates and mammals such as water voles and otters. This explains why, as in the case of the two sites described above, it is so important to undertake detailed field surveys prior to drawing up a comprehensive management plan, in order to identify the unique character of each site and to assess the relative importance of its ecological and historical features, preferably after consultation with specialists in the relevant disciplines. Indeed, the new Environmental Stewardship Scheme requires farmers to draw up a comprehensive 'Farm Environmental Plan' (FEP) before applying for the higher tier. It is hoped that these plans will help the farmer, and Natural England (formerly Defra), identify the most important areas and features in each area of land and thus provide balanced and appropriate management regimes.

However, the extent of conflicts between the different approaches to conservation should not be exaggerated. Conservation measures directed mainly towards birds, for example, can also be beneficial for the historic environment. The ditch patterns in old water meadows are often restored, and any scrub removed. Moreover, increasing soil wetness can also serve to protect buried archaeological features, by maintaining the anaerobic conditions required for the preservation of wood and other organic materials, especially in peat. Where historical considerations are a major part of the management plan, the restoration of built structures often requires careful thought, especially regarding the kinds of materials to be used, but it is not usually excessively expensive and grants of up to 80 per cent of the cost of restoration are available.

ESA and CSS thus contributed significantly to the conservation of traditional water meadows, and in some cases assisted their restoration to working condition. The new Environmental Stewardship Schemes are building on this success, highlighting priorities and reducing potential conflicts existing between different stakeholders. In practical terms, the best way forward is usually through partnerships, taking each meadow on a case-by-case basis.

CHAPTER TWELVE

Conclusion

....

Tom Williamson and Hadrian Cook

We hope that this short volume effectively and succinctly summarises the current state of knowledge on the subject of water meadows in Britain. It is certainly not the last word on 'floating' (it would be a pity if it were), and much important research remains to be carried out by historians, archaeologists and natural scientists. Indeed, although the contributors are in broad agreement over the basic issues, perceptive readers will have noted a number of areas of fruitful disagreement.

It is clear that water meadows reached their most sophisticated forms, were most widely adopted and played their most important role in the agrarian economy in the Wessex chalklands in the period between 1600 and 1850. In part their prominence here was the result of wider agricultural circumstances – the important part water meadows could play in the local form of arable, sheep-corn husbandry – but it was also because the soils, geology and geomorphology of this area made them particularly cost-effective. The Wessex meadows – most of which were bedworks, due to the form of the local valleys – evidently worked in complex ways. They warmed the soil in early spring, bringing forward grass growth by several weeks; and they encouraged, through the rather simpler process of direct irrigation, the growth of the summer hay crop. But watering also served to dress the sward with important nutrients and chemicals, and to suppress the growth of certain species and encourage others.

Outside the chalklands of Wessex floating seems to have been a significant feature of early modern agriculture mainly in the West Country and the Marcher counties – especially in Devon, Somerset, Gloucestershire, Herefordshire and Shropshire. Here, too, water meadows seem to have been used both to force an early growth of grass and to increase the hay crop, but catchworks were more commonly used than bedworks. Moreover, in most of these districts floating was perhaps valued more for the direct benefits it brought, as a way of raising livestock numbers, than for its indirect effects, of increasing manure supplies and hence arable yields. Many of these western districts, although generally well suited to specialised livestock farming (and increasingly specialising in it in the course of the fifteenth and sixteenth centuries) had always been relatively short of good-quality meadow land. Floating helped intensify livestock production.

What remains less clear is when floating in these 'classic' forms first developed. By late medieval times meadow irrigation seems to have been fairly common practice, in some parts of England at least, although whether to force an 'early bite', or merely to enhance the summer hay crop and generally improve the sward, remains uncertain: this is clearly an important area for future research. So too is the question of when floating using 'bedworks' (as opposed to catchworks) was first developed. Most pre-seventeenth-century references to floating appear to be to catchworks, or to more primitive practices such as 'floating upwards'; but it remains possible that bedwork irrigation was also known in medieval times. Field archaeologists should certainly bear this possibility in mind.

Fully developed floating, as practised in the south and west of England, evidently spread across much of the north and east of the country in the course of the later eighteenth and nineteenth centuries, and into Scotland and Wales. In most of these regions the practice seems to have had less success than in its traditional heartlands, probably because the soils, climate and other environmental circumstances were less suitable. Floating was generally adopted sparingly, and it is arguable that – in certain districts at least – it was something of a fad, encouraged by major landowners at times of particular agricultural prosperity, and perhaps in part fuelled by ideological considerations. What remains far from clear, however, is whether simple forms of meadow irrigation were already being practised in some of these areas before the adoption there of 'developed' floating, forms which – as Cummings and Cutting intimate in Chapter Seven – may have been largely ignored by elite agricultural writers like Arthur Young. The question is complicated by the fact that even in the eighteenth and nineteenth centuries meadow irrigation in these areas often seems to have been carried out for different and perhaps more limited reasons than in Wessex: to improve the quality of the hill pastures or (in the east) simply to increase the summer hay crop. Indeed, in all periods fully developed water meadows on the Wessex model, and completely unwatered meadows, should not be seen as completely separate and distinct entities but rather as two ends of a spectrum or, perhaps more accurately, as just two out of a number of ways of managing low-lying grassland.

The current condition and future management of water meadows are also matters of some interest and debate. Water meadows have an important archaeology, involving a range of built structures such as bridges, aqueducts and sluices: these are often sophisticated industrial features which serve as a powerful reminder that water meadows continued to be formed, and improved, well into the Victorian period. These, and the various kinds of earthworks associated with meadows – carriers, bedworks, drains and the rest – deserve more care and consideration than they have tended to receive in the past, although attitudes are evidently changing. Moreover, we can so easily kill what we love, and the restoration of particular examples of water meadows to working order can itself cause damage to this rich archaeology: conservation and restoration are in potential conflict. Similar problems surround the environmental impacts of restoration, for functioning water meadows are not necessarily of great value in terms of biodiversity when compared to meadows in abandoned or even derelict condition.

Natural scientists now have a fairly good idea of precisely how floating worked. Indeed, it is ironic that it is only now, after water meadows have ceased to be a major feature of our agriculture, that we can really see how they served to increase productivity. Is this new knowledge of anything other than antiquarian interest? Perhaps: yet in a world increasingly concerned with issues of 'sustainability', and in particular with the effects of large-scale agro-chemical applications, it is not altogether impossible that water meadows may one day make some kind of a comeback. Floating has been used in many different contexts in the past, with a variety of ends in view. The ability of the technique to enhance the quality and quantity of the sward without chemical inputs, and also to trap diffuse chemicals in watercourses, may yet ensure that it has a future role in agriculture and environmental management.

Abbreviations

DA	Doncaster Archives
DRO	Dorset Record Office
HRO	Hampshire Record Office
IGS	Institute of Geological Sciences
NMR	National Monuments Record
NRO	Norfolk Record Office
RCHME	Royal Commission on the Historical Monuments of England
RSPB	Royal Society for the Protection of Birds
SWA	Southern Water Authority
TNA:PRO	The National Archives: Public Record Office
WSRO	Wiltshire and Swindon Record Office
WWA	Wessex Water Authority

Bibliography

Adkin, B.W. (1933) *Land Drainage in Britain*, Estates Gazette, London.

Albone, J., Massey, S. and Tremlett, S. (2004) 'The National Mapping Programme in Norfolk, 2003–4', *Norfolk Archaeology* **44**, 549–55.

Alcock, M.B., Lovett, J.V. and Machin, D. (1968) 'Techniques used in the study of the influence of environment on primary pasture production on hill and lowland habitats', in *The Measurement of Environmental Factors in Terrestrial Ecology*, ed. R.M.Walsworth, Blackwell Scientific, Oxford, 191–203.

Amphlett, J. (1890) *A Short History of Clent*, London.

Annals of Agriculture (1803) vol. XXXIX, 322.

Arneborg, J. (2003) 'Greenland irrigation systems on a West Nordic background: an overview of the evidence of irrigation systems in Norse Greenland c. 980–1450 AD', *Ruralia* **4**: *Water Management in Medieval Rural Economy*, 137–145.

Aston, M. (1972) 'The earthworks of Bordesley Abbey, Redditch, Worcestershire', *Medieval Archaeology* **16**, 133–6.

Atwood, G. (1963) 'A study of the Wiltshire water meadows', *Wiltshire Archaeological Magazine* **58**, 403–13.

Aubrey, J., ed. K.G.Ponting (1969) *Natural History of Wiltshire*, Newton Abbot.

Ault, W.O. (1972) *Open-field Farming in Medieval England: A Study of Village By-laws*, George Allen and Unwin, London.

Bacon, R.N. (1844) *Report on the Agriculture of Norfolk*, London.

Barker, T. (1858) 'On water meadows as suitable for Wales and other mountain districts', *Journal of the Bath and West of England Society* **6**, 267–86.

Batchelor, T. (1813) *General View of the Agriculture of the County of Bedford*, Sherwood, Neely and Jones, London.

Beastall, T.W. (1978) *The Agricultural Revolution in Lincolnshire*, History of Lincolnshire Committee, Lincoln.

Bennett, H.S. (1937) *Life on the English Manor: A Study of Peasant Conditions 1150–1400*, Cambridge University Press, Cambridge.

Bettey, J.H. (1977) 'The development of water meadows in Dorset', *Agricultural History Review* **25**, 37–43.

Bettey, J.H. (1999) 'Water meadows in the southern counties of England', in *Water Management in the English Landscape*, eds H.F.Cook and T.Williamson, Edinburgh University Press, Edinburgh, 179–95.

Bettey, J.H. (2003) 'The development of water meadows in the Salisbury Avon, 1665–1690', *Agricultural History Review* **51**, 163–72.

Biddick, K. (1989) *The Other Economy: Pastoral Husbandry on a Medieval Estate*, University of California Press, Berkeley.

Binding, H. and Stevens, D. (1977) *A New History of Minehead*, The Exmoor Press, Dulverton.

Blith, W. (1653) *The English Improver Improved or the Survey of Husbandry Surveyed*, John Knight, London.

Blom, C.W.P.M. (1999) 'Adaptations to flooding stress: from plant community to molecule', *Plant Biology* **1**, 261–73.

Blom, C.W.P.M. and Voesenek, L.A.C.G. (1996) 'Flooding: the survival strategies of plants', *Trends in Ecology and Evolution* **11**, 290–5.

Boswell, G. (1779) *A Treatise on Watering Meadows: Wherein are Shewn some of the many Advantages Arising from that Mode of Practice, Particularly on Coarse, Boggy, or Barren Lands*, printed for the author, London.

Bowie, G.S. (1987) 'Water meadows in Wessex: a re-evaluation for the period 1640–1850, *Agricultural History Review* **35**, 151–8.

Bowles, N. (1988) *The Southern Wey: A Guide*, River Wey Trust, Liphook.

Boys, J. (1805; 2nd edn) *General View of the Agriculture of the County of Kent*, Richard Phillips, London.

Brady, N.C. (1974) *The Nature and Properties of Soils*, Macmillan, London.

Braudel, F. (1992) *The Mediterranean World in the Age of Philip II*, Harper Collins, London.

Brix, H. and Sorrell, B.K. (1996) 'Oxygen stress in wetland plants: comparison of de-oxygenated and reducing root environments', *Functional Ecology* **10**, 521–6.

Brown, G. (2002) *Earthworks at Buildwas Abbey, Shropshire*, English Heritage Archaeological Investigation Report Series A1/9/2002, Swindon.

Brown, G. (2003) 'Irrigation of water meadows in England', *Ruralia* **4**: *Water Management in Medieval Rural Economy*, 84–92.

Browne, J. (1817) *A Treatise on Irrigation, or The Watering of Land with some Observations on Cattle, Tillage and Planting*, London.

Burnham, B.C. and Wacher, J. (1990) *The 'Small Towns' of Roman Britain*, Batsford, London.

Burke, J.F. (1834) *British Husbandry*, Baldwin and Craddock, London.

Burton, R.A. (1989) *The Heritage of Exmoor*, Barnstable.

Caird, J. (1852) *English Agriculture 1850–51*, Frank Cass, London.

Campbell, B.M.S. (2000) *English Seigniorial Agriculture, 1250–1450*, Cambridge University Press, Cambridge.

Carrier, E.H. (1936) *The Pastoral Heritage of Britain*, Christophers, London.

Chatwin, C.P. (1960; 3rd edn) *British Regional Geology: The Hampshire Basin and Adjoining Areas*, HMSO, London.

Chibnall, A.C. (1965) *Sherington: Fiefs and Fields of a Buckinghamshire Village*, Cambridge University Press, Cambridge.

Chorley, G.P.H. (1981) 'The agricultural revolution in northern Europe, 1750–1880: nitrogen, legumes and crop production', *Economic History Review* **34**, 71–93.

Claridge, J. (1793) *General View of the Agriculture of the County of Dorset*, W. Smith, London.

Clarke, J. (1794) *General View of the Agriculture of the County of Brecknock*, J. Smeeton, London.

Clarke, J. (1795) 'On watering meadows in Brecknockshire', *Annals of Agriculture* **23**, 192–6.

Colmer, T.D. (2003) 'Aerenchyma and an inducible barrier to radial oxygen loss facilitate rot aeration in upland, paddy and deep-water rice (*Oryza sativa* L.)', *Annals of Botany* **91**, 301–9.

Cook, H.F. (1994) 'Field-scale water management in England to AD 1900', *Landscape History* **16**, 53–66.

Cook, H.F. (1999) 'Soil and water management: principles and purposes', in *Water Management in the English Landscape*, eds H.F. Cook and T. Williamson, Edinburgh University Press, Edinburgh, 15–29.

Cook, H. and Williamson, T. eds (1999) *Water Management in the English Landscape*, Edinburgh University Press, Edinburgh.

Cook, H., Stearne, K. and Williamson, T. (2003) 'The origins of water meadows in England', *Agricultural History Review* **51**, 155–62.

Cook, H.F., Cutting, R.L., Buhler, W. and Cummings, I.P.F. (2004) 'Productivity and soil nutrient relations of bedwork water meadows in southern England', *Agriculture, Ecosystems and Environment* **102**, 61–79.

Cook, H.F. and Cutting, R.L. (in preparation) 'Water temperature variations on operative bedwork water meadows'.

Correl, D. and Weller, D.E. (1989) 'Factors limiting processes in fresh water wetlands: an agricultural primary stream riparian forest', in *Freshwater Wetlands and Wildlife*, eds R.R. Sharitz and J.W. Gibbons, United States Department of Energy symposium series no 61.

Cowan, M. (1982) *Floated Water Meadows in the Salisbury Area*, South Wiltshire Industrial Archaeology Society, Monograph 9, Salisbury.

Cowan, M. (2005) *Wiltshire Water Meadows*, Hobnob Press, Salisbury.

Crawford, O.G.S. (1928) *Wessex from the Air*, Cambridge University Press, Cambridge.

Crawford, R.M.M. (1989) *The Anaerobic Retreat. Studies in Plant Survival: Ecological Case Histories of Plant Adaptation to Adversity*, Blackwell Scientific, Oxford.

Crawford, R.M.M., Jefree, C.E. and Rees, W.G. (2003) 'Paludification and forest retreat in northern oceanic environments', *Annals of Botany* **91**, 213–26.

Currie, C.K. (1998) 'Clent Hills, Worcestershire: an archaeological and historical survey', *Transactions of the Worcester Archaeological Society* **16**, 183–206.

Curtis, L.F. (1971) *The Soils of Exmoor Forest*, Soil Survey Special Survey No. 5, Harpenden.

Cushion, B. and Davison, A. (2003) *The Earthworks of Norfolk*, published as *East Anglian Archaeology* **104**, Norfolk Museums and Archaeology Service, Dereham.

Cummings, I.P.F. (1991/2) *River Nar and River Wensum Corridor Survey Report*, English Nature Reports, Bracondale, Norwich.

Cummings, I.P.F. (1992) *Some Effects of Root Competition on* Anemone nemorosa L. *in Ancient Coppiced Woodland*, unpublished PhD thesis, University of East Anglia.

Cutting, R.L. (2002) *An Investigation into the Operation and Function of Floated Water Meadows*, unpublished PhD thesis, Anglia Polytechnic University.

Cutting, R.L., Cook, H.F. and Cummings, I.P.F. (2003) 'Hydraulic conditions, oxygenation, temperatures and sediment relations of bedwork water meadows', *Hydrological Processes* **17**, 1823–43.

Cutting, R.L. and Cummings, I.P.F. (1999) 'Water meadows: their form, ecology and plant ecology', in *Water Management in the English Landscape*, eds H.F. Cook and T. Williamson, Edinburgh University Press, Edinburgh, 157–79.

Darby, H.C. (1940) *The Draining of the Fens*, Cambridge University Press, Cambridge.

Darby, J. (1873) 'The Farming of Somerset', *Journal of the Bath and West of England Society* **5**, 96–172.

Davenport, F.G. (1906) *The Economic Development of a Norfolk Manor, 1086–1565*, Cambridge University Press, Cambridge.

Davies, M.S. and Singh, A.K. (1983) 'Population and differentiation in *Festuca rubra* L. and *Agrostis stolonifera* L. in response to soil waterlogging', *New Phytologist* **94**, 573–83.

Davis, T. (1794) *General View of the Agriculture of the County of Wiltshire*, London.

Davis, T. (1811) *General View of the Agriculture of the County of Wiltshire*, Richard Phillips, London.

Defoe, D. (1724; Everyman edn, 1927) *A Tour Through the Whole Island of Great Britain*, Peter Davies, London.

Delorme, M. (1989) 'A watery paradise: Rowland Vaughan and Hereford's Golden Vale', *History Today* **39**, 38–43.

Dennison, J. (1840) 'On the Duke of Portland's water meadows at Clipstone Park', *Journal of the Royal Agricultural Society of England* **1**, 359–70.

Dilke, O.A.W. (1971) *The Roman Land Surveyors*, David and Charles, Newton Abbot.

Drew, M.C. (1997) 'Oxygen deficiency and root metabolism: injury and acclimation under hypoxia and anoxia', *Annual Review of Plant Physiology and Plant Molecular Biology* **48**, 223–50.

Dugan, S. and Dugan, D. (2000) *The Day the World Took Off: The Roots of the Industrial Revolution*, Channel Four Books/Macmillan, London.

Dunning, R.W. ed. (1974) *A History of the County of Somerset*, vol. 3, Victoria history of the counties of England, Oxford University Press for the Institute of Historical Research, Oxford.

Dunning, R.W. ed. (1978) *A History of the County of Somerset*, vol. 4, Victoria history of the counties of England, Oxford University Press for the Institute of Historical Research, Oxford.

Dunning, R.W. ed. (1985) *A History of the County of Somerset*, vol. 5, Victoria history of the counties of England, Oxford University Press for the Institute of Historical Research, Oxford.

Dunning, R.W. ed. (1992) *A History of the County of Somerset*, vol. 6, Victoria history of the counties of England, Oxford University Press for the Institute of Historical Research, Oxford.

Dyer, W.T. (1863) 'Long Grass', *Notes and Queries* 3rd series, **4**, 415.

Dyer, W.T. (1864) 'Long Grass and Water Meadows', *Notes and Queries* 3rd series, **4**, 288.

Emanuelsson, U. and Möller, J. (1990) 'Flooding in Scania', *Agricultural History Review* **38**, 127–48.

Encyclopaedia Britannica (1880; 9th edn) 'Irrigation', vol. 13, Adam and Charles Black, Edinburgh, 368–9.

Environment Agency (1988) 'Hampshire Avon local environment agency plan', consultation draft.

Everson, P. (1979) 'The reclamation of Blyton and Laughton commons', *Lincolnshire History and Archaeology* **14**, 40–2.

Exter, J. (1798) 'On irrigation, and particularly on the different effects of rich water, as a manure on different soils', *Annals of Agriculture* **30**, 204–8.

Field, J. (1993) *A History of English Field Names*, Longman, London.

Findlay, D.C., Colbourne, G.J.H., Cope, D.W., Harrod, T.R., Hogan, D.V. and Staines, S.J. (1984) *Soils and their use in South West England*, Soil Survey of England and Wales Bulletin **14**, Harpenden.

Fitzherbert, J. (1523) *The Boke of Surveying and Improvements*, Rycharde Pynson, London.

Fleming, A. (1998) *Swaledale: Valley of the Wild River*, Edinburgh University Press, Edinburgh.

Folkingham, W. (1610) *Feudigraphica: The Synopsis or Epitome of Surveying Methodized*, William Stansby for Richard Moore, London.

Forsyth, R. (1875) 'Agriculture systematically explained', in *Encyclopaedia Britannica*, 9th edn, vol. 2, Bell, Edinburgh, 362–371.

Fox, H.S.A. (1984) 'Some ecological dimensions of English medieval field systems', in *Archaeological Approaches to Medieval Europe*, ed. K. Biddick, Kalamazoo, Michigan, USA, 129–58.

Francis, P. (1984) 'The catch meadow irrigation systems found on Exmoor', unpublished BA dissertation, University of Durham.

Fraser, I. (2001) 'Three Perthshire water meadows: Strathallan, Glendevon, and Bertha', *Tayside and Fife Archaeological Journal* **7**, 133–44.

Fraser, R. (1794b) *General View of the Agriculture of Devon*, C. Macrae, London.

Garnett, W.J. (1849) 'The farming of Lancashire', *Journal of the Royal Agricultural Society of England* **10**, **1**, 1–51.

Gerard, T. (1980 edn) *Coker's Survey of Dorsetshire*, Dorset Publishing Company, Sherborne.

Gervers, M. ed. (1996) *The Cartulary of the Knights of St John of Jerusalem in England Part 2: Primera Camera, Essex*, Records of Social and Economic History New Series 23, Oxford.

Grieg, J. (1988) 'Plant resources', in *The Countryside of Medieval England*, eds G. Astill and A. Grant, Oxford University Press, Basil Blackwell, Oxford, 108–27.

Grime, J.P., Hodgson, J.G. and Hunt, R. (1990) *Comparative Plant Ecology*, Kluwer Academic Publishers.

Hall, D. (1987) *The Fenland Project*, published as *East Anglian Archaeology* **35**, Cambridgeshire Archaeological Committee in conjunction with the Fenland Project Committee and the Scole Archaeological Committee, Cambridge.

Halstead, P. (1998) 'Ask the fellows who lop the hay: leaf-fodder in the mountains of northwest Greece', *Rural History* **9**, **2**, 211–34.

Hampshire County Council (1999) *Survey of Water Meadows in Hampshire*, unpublished report, Hampshire County Council, Winchester.

Hampshire County Council (2002) *Water Meadows in Hampshire*, Hampshire County Council, Winchester.

Hampshire County Council (2003) *The Conservation of Water Meadows*, Hampshire County Council, Winchester.

Hardy, T. (1878) *Return of the Native*, London.

Harris, A. (1961) *The Rural Landscape of the East Riding of Yorkshire, 1700–1850*, Hull University Publications, London.

Hassall, C. (1794) *General View of the Agriculture of the County of Pembroke*, J. Smeeton, London.

Heddle, R.G. and Ogg, W.G. (1933) 'Experiments in the improvement of hill pasture', *Scottish Journal of Agriculture* 16, 431–46.

Hill, M.O. (1993) *TABLEFIT Version 0.0 for Identification of Vegetation Types*, Institute of Terrestrial Ecology, Huntingdon.

Hodge, C., Burton, R., Corbett, W., Evans, R. and Searle, R. (1984) *Soils and their Use in Eastern England*, Soil Survey of England and Wales, Harpenden.

Hunn, J. (1994) *Reconstruction and Measurement of Landscape Change*, British Archaeological Reports 236, Oxford.

Institute of Geological Sciences (1981) *Hydrogeological Map of the Northern East Midlands*, scale 1:100,000, London.

Institute of Geological Sciences/Southern Water Authority (1979) *Hydrogeological Map of Hampshire and the Isle of Wight*, scale 1:100,000, London.

Institute of Geological Sciences/ Wessex Water Authority (1979) *Hydrogeological Map of the Chalk and Associated Minor Aquifers of Wessex*, scale 1:100,000, London.

James, J.F. and Bettey, J.H. (1993) *Farming in Dorset: The Diary of James Warne 1758 and the Letters of George Boswell 1787–1805*, Dorset Record Society, Dorchester.

Jamieson, E. (2001) *Survey of Larkbarrow Farm, Exmoor, Somerset*, unpublished report, English Heritage, London.

Johnson, J.S. (1990) *Chester's Roman Fort*, English Heritage, London.

Jones, G., Blakey, I. and MacPherson, E. (1960–2) 'Dolaucothi: the Roman aqueduct', *Bulletin of the Board of Celtic Studies* 19, 71–9.

Justin, S.H.F.W. and Armstrong, W. (1987) 'The anatomical characteristics of roots and plant responses to soil flooding', *New Phytologist* 106, 465–95.

Kerridge, E. (1953a) *Surveys of the Manors of Philip, First Earl of Pembroke*, Wiltshire Archaeological and Natural History Society, Devizes.

Kerridge, E. (1953b) 'The floating of the Wiltshire water meadows', *Wiltshire Archaeological Magazine* 55, 105–18.

Kerridge, E. (1954) 'The sheepfold in Wiltshire and the floating of the watermeadows', *Economic History Review* 6, 282–9.

Kerridge, E. (1956) 'Turnip husbandry in High Suffolk', *Economic History Review* 2nd Ser. 8, 390–2.

Kerridge, E. (1967) *The Agricultural Revolution*, George Allen and Unwin, London.

Kerridge, E. (1973) *The Farmers of Old England*, George Allen and Unwin, London.

Kerridge, E. (1992) *The Common Fields Of England*, Manchester University Press, Manchester.

Kramer, P.J. and Boyer, J.S. (1995) *Water Relations in Plants and Soils*, Academic Press, London.

Lamb, H.H. (1995; 2nd edn) *Climate, History and the Modern World*, Routledge, London and New York.

Larking, L.B. and Kemble, J.M. eds (1857) *The Knights Hospitallers in England: Being the Report of Prior Philip de Thame to the Grand Master, Elyan de Villanova, for AD 1338*, Camden Society, London.

Lillie, M. (1998) 'Alluvium and warping in the lower Trent Valley', in *Wetland Heritage of the Ancholme and Lower Trent Valleys*, eds R. Van de Noort and G. Ellis, Humber Wetland Project, University of Hull, 102–22.

MacDonald, J. (1908; 5th edn) *Stephens' Book of the Farm*, vol. 1, William Blackwood, London.

Manning, O. and Bray, W. (1804) *The History and Antiquities of the County of Surrey*, vol. 1, J. White, London.

Marsh, T.J. and Lees, M.L. (2003) *Hydrological Data United Kingdom, Hydrometric Register and Statistics 1996–2000*, Natural Environment Research Council, Centre for Ecology and Ecology, Wallingford, Oxford.

Marshall, W. (1788) *The Rural Economy of Yorkshire*, vol. 1, T. Cadell, London.

Marshall, W. (1796a) *The Rural Economy of the West of England*, vol. 1, G. Nicol, London.

Marshall, W. (1796b) *The Rural Economy of the West of England*, vol. 2, G. Nicol, London.

Marshall, W. (1796c) *The Rural Economy of the Midland Counties*, vol. 1, G. Nicol, London.

McKinley, J. (2003) 'A Wiltshire "Bog Body"?', *Wiltshire Archaeological Magazine* 96, 7–18.

McOmish, D., Field, D. and Brown, G. (2002) *The Field Archaeology of the Salisbury Plain Training Area*, English Heritage, Swindon.

Mignot, P. and De Meulemeester, J. (2003) 'A propos de l'hydraulique en Ardenne belge', *Ruralia* 4: *Water Management in Medieval Rural Economy*, 3–10.

Mingay, G.E. ed. (1977) *Documents in Economic History: The Agricultural Revolution*, Adam and Charles Black, London.

Monkhouse, R.A. and Richards, H.J. (1982) *Groundwater Resources of the United Kingdom*, Commission of the European Communities, Hanover.

Moon, H.P. and Green, F.H.W. (1940) 'Water meadows in Southern England', in *The Land of Britain, The Report of the Land Utilisation Survey of Britain*, ed. L.Dudley Stamp, pt. 89, Appendix 2, Geographical Publications Ltd, London, 373–90.

Moorhouse, S.A. (1981) 'Water supply', in *West Yorkshire: An Archaeological Survey to AD 1500*, eds M.L.Faull and S.A.Moorhouse, 3 vols, West Yorkshire Metropolitan County Council, Wakefield, 695–9.

Moss, B. (1988; 2nd edn) *The Ecology of Fresh Waters*, Blackwell Scientific, Oxford.

Muir, R. (2000) 'Pollards in Nidderdale: a landscape history', *Rural History* 11, 1, 95–111.

Newton, K.C. (1960) *Thaxted in the Fourteenth Century*, Essex Records Office, Chelmsford.

Northcote, S.H. (1855) 'A few words on water-meadows', *Journal of the Bath and West of England Society* 3, 112–16.

Ojala, A., Kankaala, P. and Tulonen, T. (2001) 'Growth response of *Equisetum fluviatile* to elevated CO_2 and temperature', *Environmental and Experimental Botany* 47 (2), 157–71.

Orwin, C.S. (1929) *The Reclamation of Exmoor Forest*, Oxford University Press, London.

Orwin, C.S. and Orwin, C.S. (1938) *The Open Fields*, Clarendon Press, Oxford.

Page, C.N. (1997; 2nd edn) *The Ferns of Britain and Ireland*, Cambridge University Press, Cambridge.

Palliser, D.M. (1976) *The Staffordshire Landscape*, Hodder and Stoughton, London.

Parkinson, R. (1813) *General View of the Agriculture of the County of Huntingdon*, Sherwood, Neely and Jones, London.

Paxton, W. (1840) 'Practical statement on the formation of an economically important water meadow', *Journal of the Royal Agricultural Society of England* 1, 346–8.

Peacock, J.M. (1975) 'Temperature and leaf growth in *Lolium perenne* 1. The thermal micro-climate: its measurement and relation to crop growth', *Journal of Applied Ecology* 12, 99–114.

Pearce, W. (1794) *General View of the Agriculture of the County of Berkshire*, W.Bulmer, London.

Perry, P.J. (1974) *British Farming in The Great Depression, 1870–1914*, David and Charles, Newton Abbot.

Philips, C.W. ed. (1970) *The Fenland in Roman Times*, Royal Geographical Society Research Series 5, London.

Pitt, W. (1795) 'On watering meadows', *Annals of Agriculture* XXIII, 539–44.

Plumb, J.H. (1952) 'Sir Robert Walpole and Norfolk husbandry', *Economic History Review* 2nd Ser. 5, 86–9.

Pusey, P. (1845) Editor's note after Mr Roal's article, 'On converting a Moory Hillside into a Catch Meadow', *Journal of the Royal Agricultural Society of England* 6, 521.

Pusey, P. (1849) 'On the theory and practice of water meadows', *Journal of the Royal Agricultural Society of England* 10, 462–79.

Rackham, O. (1976) *Trees and Woodland in the British Landscape*, Dent, London.

Rackham, O. (1986) *The History of the Countryside*, Dent, London.

RCHME (1952–75) *Dorset*, vols 1–6, HMSO, London.

Read, C.S. (1849) 'The farming of South Wales', *Journal of the Royal Agricultural Society of England* 10, 1, 122–65.

Reed, M. (1979) *The Buckinghamshire Landscape*, Hodder and Stoughton, London.

Richards, K. (1982) *Rivers: Form and Process in Alluvial Channels*, Methuen, London.

Richardson, R.C. (1984) 'Metropolitan counties: Bedfordshire, Hertfordshire, and Middlesex', in *The Agrarian History of England and Wales, Volume V: 1640–1750 Part 1: Regional Farming Systems*, ed. J.Thirsk, Cambridge University Press, Cambridge, 239–69.

Richmond, I. (1940–1) 'The water supply of the Roman fort at Lyne, Peebleshire', *Proceedings of the Society of Antiquaries of Scotland* 75, 39–43.

Riley, H. and Wilson-North, R. (2001) *The Field Archaeology of Exmoor*, English Heritage, Swindon.

Roal, J. (1846) 'On the converting of mossy hillside to catch meadow', *Journal of the Royal Agricultural Society of England* 6, 518–22.

Robertson, J. (1794) *General View of the Agriculture in the Southern Districts of the County of Perth*, J.Nichols, London.

Robie, D. (1876) 'On English water meadows and how far they are applicable to Scotland', *Transactions of the Highland and Agricultural Society*, Series 4, 4, 87–97.

Robinson, D.H. (1949; 13th edn) *Fream's Elements of Agriculture*, John Murray, London.

Rodwell, J.S. ed. (1991) *British Plant Communities, Volume 2: Mires and Heath*, Cambridge University Press, Cambridge.

Rodwell, J.S. ed. (1993) *British Plant Communities, Volume 3: Grassland and Montane Communities*, Cambridge University Press, Cambridge.

Rodwell, J.S. ed. (1995) *British Plant Communities, Volume 4: Aquatic Communities, Swamps and Tall-herb Fens*, Cambridge University Press, Cambridge.

Royal Society for the Protection of Birds, English Nature and Institute of Terrestrial Ecology (1997) *The Wet Grassland Guide: Managing Floodplain and Coastal Wet Grassland for Wildlife*, RSPB, Sandy.

Rozema, J. and Blom, B. (1977) 'Effects of Salinity and Inundation on the Growth of *Agrostis stolonifera* and *Juncus gerardii*', *The Journal of Ecology* 65, 1, 213–22.

Russell, E.W. (1973; 10th edn) *Soil Conditions and Plant Growth*, Longmans, London.

Saunders, H.W. (1916) 'Estate Management at Raynham 1661–86 and 1706', *Norfolk Archaeology* **19**, 39–67.

Schmaedecke, M. (2003) 'Beobachtungen zur Wassernutzung auf dem Lande wahrend des Mittelalters und der fruhen Neuzeit im Gebiet der heutigen Schweiz', *Ruralia* **4**: *Water Management in Medieval Rural Economy*, 188–98.

Scott, J. (1883) *Irrigation and Water-Supply: A Practical Treatise on Water Meadows, Sewage Irrigation and Warping*, C. Lockwood and Co., London.

Sealy, K. R. (1955) 'The terraces of the Salisbury Avon', *Geographical Journal* **121**, 350–6.

Seymour, R. (1796) 'Experiments in irrigation and other objects of rural economy', *Annals of Agriculture* **26**, 447–83.

Sheail, J. (1971) 'The formation and maintenance of water meadows in Hampshire, England', *Biological Conservation*, **3**, 2, 101–06.

Sheail, J. (1996) 'Town wastes, agricultural sustainability and Victorian sewage', *Urban History* **23**, 2, 189–210.

Sheil, J. (1991) 'Soil Fertility in the Pre-Fertiliser Era', in *Land, Labour and Livestock*, eds B. Campbell and M. Overton, Manchester University Press, Manchester, 51–77.

Simmons, B. (1979) 'The Lincolnshire Car Dyke', *Britannia* **10**, 183–90.

Smith, J. (1799) 'On the advantages of watering pasture and meadow grounds in the Highlands', *Transactions of the Highland Agricultural Society*, Series 1, 1.

Smith, R. (1851) 'Some account of the formation of hillside catch-meadows on Exmoor', *Journal of the Royal Agricultural Society of England* **12**, 139–48.

Smith, R. (1856a) 'Irrigation-temperature of springs', *Journal of the Bath and West of England Society* **4**, 295–7.

Smith, R. (1856b) 'Bringing moorland into cultivation', *Journal of the Royal Agricultural Society* **17**, 349–94.

Smith, W. (1806) *Observations on the Utility, Form and Management of Water Meadows, and the Draining and Irrigating of Peat Bogs*, R. M. Bacon, Norwich.

Spedding, C. R. W. and Diekmahns, E. C. (1972) *Grasses and Legumes in British Agriculture*, Bulletin of the Commonwealth Bureau of Pastures and Field Crops 49, Farnham.

Stace, C. (1992) *New Flora of the British Isles*, Cambridge University Press, Cambridge.

Stamp, L. Dudley (1950) *The Land of Britain: its use and misuse*, Longmans, Green and Co, London.

Stearne, K. I. (2004) *Water Meadows in the English Landscape: Conflict, Compromise and Change*, unpublished PhD, University of London.

Stearne, K., Cook, H. and Stearne, P. (2002) 'Water meadows in southern England', *Landwards (Journal of Institute of Agricultural Engineers)* **57**, **5**, 2–6.

Steele, N. and Tatton-Brown, T. (n.d.) *The History of the Harnham Water Meadows*, Harnham Water Meadows Trust, Salisbury.

Stephens, G. (1834) *The Practical Irrigator. Being an Account of the Utility, Formation and Management of Irrigated Meadows, with a Particular Account of the Success of Irrigation in Scotland*, R. Miller, Edinburgh.

Stevenson, W. (1815) *General View of the Agriculture of the County of Dorset*, Sherwood, Neely and Jones, London.

Street, A. (1932) *Farmer's Glory*, Faber and Faber, London.

Street, A. (1946) *Round the Year on the Farm*, Faber and Faber, London.

Subbaiah, C.C. and Sachs, M.M. (2003) 'Molecular and cellular adaptations of maize to flooding stress', *Annals of Botany* **91**, 119–27.

Taylor, C. (1973) *The Making of the Cambridgeshire Landscape*, Hodder and Stoughton, London.

Taylor, C. (1974) 'Total archaeology', in *Landscapes and Documents*, eds A. Rogers and T. Rowley, Standing Conference on Local History, London, 15–26.

Taylor, C. (1999) 'Post-medieval drainage', in *Water Management in the English Landscape*, eds H. F. Cook and T. Williamson, Edinburgh University Press, Edinburgh, 141–56.

Taylor, C. (2002) 'Seventeenth-century water-meadows at Babraham', *Proceedings of the Cambridge Antiquarian Society* **91**, 103–17.

Thirsk, J. (1987) *England's Agricultural Regions and Agrarian History 1500–1750*, Macmillan, London.

Thomas, S. (1998) 'Forgotten harvests: the history and conservation of water meadows', *British Wildlife* **10**, 82–8.

Thomasson, A.J. and Youngs, E.G. (1975) 'Water movement in soil', *Technical Bulletin of the Ministry of Agriculture, Fisheries and Food* No. **29**.

Trimmer, C. P. (1969) 'The turnip, the new husbandry and the English agricultural revolution', *Quarterly Journal of Economics* **83**, 375–95.

Trought, M.C.T. and Drew, M.C. (1980) 'The development of waterlogging damage in young wheat plants in anaerobic solution cultures', *Journal of Experimental Botany* **31**, 1573–85.

Turner, M.E. (1982) *Volume 190: Home Office Acreage Returns HO67. List and Analysis*, 3 parts, London.

Turner, S. (2004) 'The changing ancient landscape: south-west England c. 1700–1900', *Landscapes* **5.1**, 18–34.

Vancouver, C. (1813) *General View of the Agriculture of Hampshire*, Sherwood, Neely and Jones, London.

van Beers, W.F.J. (1958) *The auger hole method, a field measurement of the hydraulic conductivity of soil below the water table*, International Institute for Land Reclamation and Improvement, Bulletin No. 1.

van Elburg, H., Engelen, G.B. and Hemker, C.J. (1989) *FLOWNET version 5.12. User's Manual.* Institute of Earth Sciences at the Free University, Amsterdam.

Vaughan, R. (1610; 1897 edn) *The Most Approved and Long Experienced Water Workes*, republished with preface by Ellen Beatrice Wood, John Hodges, London.

Visser, E.J.W., Colmer, T.D., Blom, C.W.P.M. and Voesenek, L.A.C.G. (2000) 'Changes in growth, porosity, and radial oxygen loss from adventitious roots of selected mono- and dicotyledenous wetland species with contrasting types of aerenchyma', *Plant, Cell and Environment* **23** (**11**), 1237–46.

Visser, E.J.W., Nabben, R.H.M., Blom, C.W.P.M. and Voesenek, L.A.C.G. (1997) 'Elongation by primary lateral roots and adventitious roots during conditions of hypoxia and high ethylene concentrations', *Plant, Cell and Environment* **20**, 647–53.

Visser, E.J.W., Voesenek, L.A.C.G., Vartepetian, B.B. and Jackson, M.B. (2003) 'Flooding and Plant Growth', *Annals of Botany* **91**, 107–9.

Wade Martins, S. (1995) *Fields and Farms*, Batsford, London.

Wade Martins, S. and Williamson, T. (1994) 'Floated water-meadows in Norfolk: a misplaced innovation', *Agricultural History Review* **42.1**, 20–37.

Wade Martins, S. and Williamson, T. (1999a) *Roots of Change: Farming and the Landscape in East Anglia 1700–1870*, British Agricultural History Society Monograph, Exeter.

Wade Martins, S. and Williamson, T. (1999b) 'Inappropriate technology? The history of floating in the north and east of England', in *Water Management in the English Landscape*, eds H.F.Cook and T.Williamson, Edinburgh University Press, Edinburgh, 196–209.

Wellings, S.R. and Cooper, J.D. (1983) 'The variability of recharge of the English chalk aquifer', *Agricultural Water Management* **6**, 243–53.

White, R.E. (1997) *Principles and Practice of Soil Science: The Soil as a Natural Resource*, Blackwell Scientific, Oxford.

Whitehead, J. (1967) 'The management and land-use of water meadows in the Frome valley', *Proceedings of the Dorset Natural History and Archaeological Society* **89**, 257–81.

Whitehead, J. (1981) 'North Meadow, Cricklade', *Wiltshire Archaeological Magazine* **76**, 129–40.

Whitelock, D. (1955) *English Historical Documents: Volume 1, 500–1042*, Eyre and Spottiswoode, London.

Wild, A. ed. (1988; 11th edn) *Russell's Soil Conditions and Plant Growth*, Longman Scientific and Technical, London.

Williams, M. (1970) *The Draining of the Somerset Levels*, Cambridge University Press, Cambridge.

Williamson, T. (2003) *Shaping Medieval Landscapes: Settlement, Society, Environment*, Windgather Press, Bollington.

Wilson, R.C.L., Drury, S.A. and Chapman, J.L. (2000) *The Great Ice Age*, Routledge for the Open University, London.

Wimpey, J. (1786) 'Answers to queries respecting watered meadows', *Annals of Agriculture* **V**, 292–8.

Wordie, J.R. (1984) 'The South: Oxfordshire, Buckinghamshire, Berkshire, Wiltshire, and Hampshire', in *The Agrarian History of England and Wales, Volume V: 1640–1750 Part 1: Regional Farming Systems*, ed. J.Thirsk, Cambridge University Press, Cambridge, 317–57.

Worlidge, J. (1669; 4th edn) *Systema Agriculturae*, Tho. Dring, London.

Wright, Rev. T. (1789) *An Account of the Advantages and Methods of Watering Meadows by Art*, S.Rudder, Cirencester.

Wright, T. (1790; 2nd edn) *The Advantages and Method of Watering Meadows By Art*, S.Rudder, Cirencester.

Wright, T. (1799) *The Art of Floating Land, as is Practised in the County of Gloucester*, J.Rider, London.

Wykes, D.L. (2004) 'Robert Bakewell of Dishley', *Agricultural History Review* **52**, 38–55.

Young, A. (1770) *A Six Months Tour through the North of England*, 3 vols, W.Strahan, London.

Young, A. (1771) *A Six Weeks Tour through the Southern Counties of England and Wales*, J.Milliken, Dublin.

Young, A. (1796) *Annals of Agriculture*, **XVI**.

Young, A. (1804a) *General View of the Agriculture of the County of Norfolk*, R.Phillips, London.

Young, A. (1804b) *General View of the Agriculture of the County of Hertfordshire*, G. and W.Nicol, London.

Young, A. (1813a) *General View of the Agriculture of the County of Lincolnshire*, London.

Young, A. (1813b) *General View of the Agriculture of the County of Suffolk*, Sherwood, Neely and Jones, London.

Index

Page numbers in **bold** refer to the Figures